Modern Science and
Human Values

Modern Science and Human Values

WILLIAM W. LOWRANCE

The Rockefeller University

New York • *Oxford*
OXFORD UNIVERSITY PRESS

Oxford University Press
Oxford London New York Toronto
Delhi Bombay Calcutta Madras Karachi
Petaling Jayas Singapore Hong Kong Tokyo
Nairobi Dar es Salaam Cape Town
Melbourne Auckland

and associated companies in
Beirut Berlin Ibadan Mexico City Nicosia

Selections from works by the following authors were made possible by the kind permission of their respective publishers and representatives:

Renée C. Fox: *Essays in Medical Sociology,* John Wiley & Sons, Inc. (1979), p. 468. Reprinted by permission of the publisher.

Arthur W. Galston: "Science and social responsibility: A case history," *Annals of the New York Academy of Sciences* 196 (1972), p. 223. Reprinted by permission.

Mel Horwitch: *Clipped Wings: The American SST Conflict,* MIT Press, Cambridge, Mass. (1982), p. 336. Copyright © 1982 by The Massachussetts Institute of Technology.

National Academy Press: *Food Safety Policy: Scientific and Societal Considerations* (1979); *Environmental Impact of Stratospheric Flight* (1975); *Scientific Communication and National Security* (1982); *The National Halothane Study* (1969); *The National Research Council in 1977* (1978); *News Report* (1969); *The National Research Council in 1978* (1979); and the Resolution passed by the National Academy of Sciences on April 28, 1981, all reprinted with permission of the National Academy Press, Washington, D.C.

Robert Nisbet: "Knowledge Dethroned," *The New York Times Magazine,* Sept. 28, 1975. Copyright © 1975 by The New York Times Company. Reprinted by permission.

Mark Siegler: "Confidentiality in medicine—a decrepit concept," *New England Journal of Medicine* 307 (1982), p. 1519, Copyright © 1982 by the Massachussets Medical Society. Reprinted by permission.

Robert M. Veatch: "Justice and valuing lives," in Steven E. Rhoads, ed. *Valuing Life: Public Policy Dilemmas,* Westview Press, Boulder, Colo. (1980), p. 158. Used by permission of Westview Press.

Library of Congress Cataloging in Publication Data

Lowrance, William W., 1943-
    Modern science and human values.

    Bibliography: p.
    Includes index.
    1. Science—Social aspects. 2. Technology—Social aspects. 3. Humanities. I. Title.
Q175.5.L68 1985   303.4'83   84-29609
ISBN 0-19-503605-0
ISBN 0-19-504211-5 (pbk)

Printing (last digit): 9 8 7 6 5 4 3 2 1

Printed in the United States of America

To Joshua Lederberg,
who gave me the opportunity
to attempt it

Acknowledgments

Just as it was a superb institution in which to pursue my graduate scientific education a few years back, The Rockefeller University was a fine haven in which to pursue this project. I am deeply pleased to dedicate the book to Joshua Lederberg, the president of the university, who, without knowing how the book would turn out, encouraged me in the task and provided me with the appointment and opportunity in which to write it. Throughout the four years it took to develop this book I have been fortunate in having the able and patient research assistance of Meta Wyss, who was a devoted partner in the work. My thanks also go to Sonya Mirsky and her staff of the University Library. I very much appreciate the hard work by, first, Karen Johnson, and then, through countless finicky revisions, Susan Sheridan, to process these words into text.

Midway through the writing I benefited, as so many scholars have, from a month's fellowship at the Bellagio Study Center. For this reflective interlude I am indebted to the sponsoring Rockefeller Foundation, and to the guiding spirits of Villa Serbelloni, Roberto and Gianna Celli.

This research and writing was pursued as part of the Life Sciences and Public Policy Program of The Rockefeller University. Support for the program was provided by a series of unrestricted small grants, for which I am extremely grateful, from the American Industrial Health Council, The Bristol-Myers Fund, Chevron U.S.A., Diamond Shamrock Corporation, The Dow Chemical Company, E.I. du Pont de Nemours & Company, Exxon Education Foundation, Occidental Chemical Corporation, Mobil Foundation, Monsanto Company, The Pfizer Foundation, PPG Industries Foundation, Shell Companies Foundation, Stauffer Chemical Company, and Union Carbide Corporation.

That there still is a community of scholars, I can confirm appreciatively, from my experience with the generosity of the many experts, quite a few of whom I have never met in person, who responded to my inquiries. Here I would express my special gratitude, for their commenting on manuscript material and for other reasons known to them, to Joel Cohen, Samuel Florman, Rodney Nichols, Helen Samuels, Burton Singer, and Richelle Spindel.

New York   W.W.L.
January 1985

Contents

Introduction

Were this an earlier day when long descriptive titles were customary, this book would be called something like *Science, Technology, Medicine, and the Social Sciences: An Explication of Their Influence on Human Values, and Vice Versa; With Special Emphasis on Connecting Theory to Pragmatics*.

Science influences human values, and human values influence science. These are commonplace assertions. But underlying them a host of questions beg to be explored about what "science" and "technology" consist in, what their fundamental social ramifications are, what "human values" are and how we know our own and others', how values and science intersect, and how social values can be brought to bear on complex technical enterprises. Mapping the themes of that exploration is the task of this book.

It is my conviction that science *can* be oriented in service of human values. But to achieve this more fully than at present will require that we apprehend the essence of sociotechnical problems in much more subtle and powerful ways than heretofore. It will require that we make sustained, deliberate efforts to couple science and technology with values. And it will require that we cultivate stewardship, in both senses of that venerable word: nourishing the vitality and integrity of science, and dedicating science to the fulfillment of needs and the advancement of culture.

This book germinated in my earlier studies of public decisions about health, environmental, and nuclear-weapons-proliferation risks, in which questions continually arose about distinctions between facts and values, about the value-ladenness of professional judgment, about scientists' advising, whistleblowing, and social responsibility in general, about the ethicality of such methods as risk–benefit analysis, about political controls on technology, and about a cluster of perennial concerns in social philosophy. Inevitably these themes came to pervade this book as well.

Reaching beyond that, though, I took the opportunity to broadly survey the physical and biological sciences, the social sciences, medicine, and engineering. The comparisons yield rewarding insights. Moreover, despite the peculiarities and partitions among all these fields and their subfields, the interactions among them are profound. In today's struggle to provide sustenance for the world's hungry poor, for example, solutions are being pursued jointly in the sciences of plant genetics and pest control, in nutrition and other medical and public health research, in hydrological, transportation, and refrigeration engineering, and in social-science research ranging from microeconomics to community anthropology. Such complementarities are not unusual. The broad view reveals many opportunities for reevaluation, reform, and guidance.

Central among my purposes is providing a succinct but vivid conspectus of the public aspects of technical work. Although this will not teach my scientific colleagues much they don't already know about their own fields, it may introduce specialists to aspects of fields different from their own, provide telling comparisons, and communicate the vitality of work on these hybrid problems. Most important, I hope the depictions will serve as reference formulations, to be discussed and refined. Too often people complain about, say, science and responsibility without being clear on what "science" and "responsibility" are. Much of the pertinent literature is weak. As this manuscript grew I found myself having to construct critiques of aspects of science itself, of technology, engineering, medicine, and the social sciences, and of such guiding concepts as values, professionalism, responsibility, scientific community, public interest, and freedom of inquiry.

A few notes about the text. To be concrete I will mention many examples, often to illustrate more than just the point at hand. To convey tone of discourse I will quote protagonists and commentators in their own voices. To avoid such cumbersome modifying phrases as "scientific, technological, and medical," often I will use the generic "technical," as in "technical community"; wherever necessary, I will be more specific. "Social" means "among people," whereas "societal" means "having to do with organized society, with structured, purposively interacting sets of people." "He" and "man" refer to both genders. Those sources of so many of my examples, the National Academy of Sciences and the National Research Council, are abbreviated NAS and NRC. The bibliography, arranged alphabetically, is at the end of the book. Longer references are footnoted; short citations are parenthesized in the text (Gould, 1981, p. 15), with the name of the author or authoring institution omitted if given in the sentence. Some citations to revised or reprinted editions show the date of original publication in brackets for historical interest. Unless stated otherwise, words italicized in quoted passages were emphasized in the original.

The book will proceed as follows.

CHAPTER ONE, THE RELATION OF SCIENCE AND TECHNOLOGY TO HUMAN VALUES, raises the basic issues. In a preliminary way it probes the meanings of "values" and "value." Then it cites a variety of ways science and technology influence social philosophies and choices. As a prevading perspective, it sets out a view of technical progress as directed tragedy.

CHAPTER TWO, THE CHANGING SOCIOTECHNICAL MILIEU, describes recent trends toward politicization and politicking in technical endeavors. It then surveys the attributes of modern technology, portrays the ways technologies serve as instruments of authority and power, and discusses the tendency toward "medicalization" of life.

CHAPTER THREE, THE COMPLEXION OF SCIENCE AND THE SOCIAL SCIENCES, examines the nature of science-knowing, the internal norms of science, and such notions as "the scientific method" and "objectivity." It goes on to survey the attributes of the social sciences (anthropology, economics, and the like), and to call attention to their contributions in studying people's values and in providing reflexivity for the larger technical enterprise.

CHAPTER FOUR, PROFESSIONALISM AND RESPONSIBILITY, critiques professionalism and professional judgment. After exploring the limits of the idea of responsibility, it concludes that what society needs from technical experts is stewardship that extends far beyond traditionally defined responsibility.

CHAPTER FIVE, THE ARCHITECTONICS OF TECHNICAL TRUST, describes a triad of interlocking essential elements: technical communities (the core operative units); bargainable compacts between those technical communities and society (the dynamic frameworks); and peer review (the crucial quality assurance mechanism). It then replays the recombinant DNA controversy of the 1970s as an example of compact renegotiation.

CHAPTER SIX, SOCIETAL GUIDANCE OF INQUIRY AND APPLICATION, discusses the possibilities and limits of scientific freedom. Then it surveys the many ways society exerts guidance over the objectives, means, and procedures of technical activity.

CHAPTER SEVEN, SYSTEMATIC ASSESSMENT FOR DECISIONMAKING, portraying present times as the Age of Studies, reviews analytic concepts, such as risk, cost, efficacy, benefit, and cost-effectiveness, that can be used in assessing and anticipating social impacts of science and technology. It makes suggestions for dealing with fact/value distinctions and with factual uncertainty and indeterminacy. The chapter concludes by recounting the example of the Rasmussen study of nuclear power risks.

CHAPTER EIGHT, TAKING ACCOUNT OF SOCIAL VALUES AND ETHICS, develops more fully the concepts of "values" and "value," and examines ways they can be analyzed and considered in decisionmaking. It critiques some philosophical underpinnings of social ethics. Then it discusses several kinds of ethical challenge that are arising, regarding desirability of "the natural," obligations to future generations, paternalism, and protective discrimination.

CHAPTER NINE, SCIENCE AND TECHNOLOGY IN THE *POLIS*, draws together the themes of advising, advocating, and whistleblowing that have been running through all the preceding chapters. It points out difficulties with conceptions of "public goods," "the public interest," and "public participation." The chapter reviews ways of managing and reducing undue polarization and acrimony in sociotechnical disputes. Then it discusses the related needs for structuring public agendas.

CHAPTER TEN, STEWARDSHIP BEYOND NARROW RESPONSIBILITY, brings the book to conclusion by suggesting needs and opportunities for technical stewardship.

Modern Science and
Human Values

One

The Relation of Science and Technology to Human Values

Admiration of the extraordinary powers of science often tempts people to hope that the laboratory and clinic will hand down social *oughts*, "Thou shalts." Convictions on the issue range from the view that science and technology can and must be used to generate human values, to the view that science and technology are just as value-neutral as banking or playing soccer are and have little moral or political character.

Neither extreme, I will argue, is correct. Although scientific knowledge, once attained, may be considered *ambi-potent* for good or evil, the work of pursuing new science and developing technologies is by no means value-neutral (as, in the context of international politics, banking and playing soccer may not be, either). And although they don't dictate values, technical analyses and accomplishments profoundly influence social philosophies and choices.

VALUES AND FACTS: THE BASIC RELATIONS

Does science generate moral oughts? Almost, on occasion, but usually not, and never without reference to embedded social values. Observations of unfitness in offspring of incestuous human mating, reinforced by analogous observations in nonhuman species and long proscription by most societies as a "crime against nature," support the almost universal taboo against incest. Technical estimates of the physical, biological, and social consequences of nuclear war make us dread it. But even in these extreme cases our responses still depend on value judgments that lie outside science: abhorrence of giving birth to defective children, abhorrence of genocide.[1]

Toward desired ends, enabling oughts can be formulated in the light of knowledge from many sources, including—powerfully—science. As we become aware of environmental connections and consequences, we weave ecological sensibilities into our values fabric. As we accumulate evidence on how life-habits affect personal health, we reassign social responsibilities for health promotion.

Social values simply cannot be derived from science qua science alone. To speak, as some cavalierly do, of "the values of science" may mislead. To be sure, over the past several centuries scientists have developed tenets of method, evidence, and proof, and they have cultivated an ethos of intellectual openness, truthfulness, and international fraternity. Some writers have been so impressed with the ethics and etiquette through which scientific work proceeds that they have urged that these mores be adopted as the foundation of social ethics. In *Science and the Social Order* Bernard Barber suggested that the "rationality, universalism, individualism, 'communality,' and 'disinterestedness'" that serve science so effectively "could even some day become the dominant moral values for the whole society" (1952, p. 90). Anatol Rapoport went so far as to say that "the ethics of science must become *the* ethics of humanity" (1957, p. 798). Jacob Bronowski argued that since science flourishes in societies fostering such "values of science" as "independence and originality, dissent and freedom and tolerance," such norms should be adopted for other social endeavors as well (1965, p. 62). But science has no monopoly on creativity, truthtelling, or tolerance, nor is it uniquely the definer of these traits. Science's precedent is hardly a sufficient model for the redesign of social ethics.

Technical people contribute richly to the alleviation of suffering and the enhancement of culture. Like everyone else, scientists hold deeply cherished personal convictions, which they express often and articulately. Groups of scientists, very large groups even, vigorously pursue social goals. Coalitions may go so far as to engage in partisan politicking, as the Scientists, Engineers, and Physicians for Johnson/Humphrey did in the 1964 presidential campaign. But such actions derive no more from the methods or scientific knowledge of Pasteur or Bohr than the 1983 Artists Call Against U.S. Intervention in Central America derived from the aesthetic tenets or *oeuvre* of Turner or Cézanne.

On the other hand, any assertion that scientific activity is value-free or value-neutral is disingenuous. Disclaimers have been made at least since Robert Hooke's 1663 proposed charter for the Royal Society: "The business and design of the Royal Society is—To improve the knowledge of

1. Edel, 1955, and Margenau, 1964, are typical but disappointing earlier efforts to strengthen the conjunction between science and values. Neither book removes the contingency of oughts on what Margenau called "primary value postulates," such as "respect for human life."

naturall things, and of all useful Arts, Manufactures, Mechanick Practises, Engynes and Inventions by Experiments—(not meddling with Divinity, Metaphysics, Moralls, Politicks, Grammar, Rhetorick or Logick)" (Lyons, 1944, p. 41). But even as Hooke drafted that antiseptic mandate, interpretations of Divinity and Metaphysics were brought under severe challenge by science; Mechanick Practises, Engynes, and Inventions were pursued that would profoundly affect the Moralls of warfare and Politicks of labor; studies of the Grammar of the world's languages grew through philology toward modern semantics; and lines of Rhetorick and Logick were explored that would lead straight to G. E. Moore and Bertrand Russell.

When, in London in 1838, William Whewell, Charles Babbage, and their colleagues founded the *Journal of the Statistical Society* they chose as its symbol the wheatsheaf, to stand for the facts the journal would gather that "alone can form the basis of correct conclusions with respect to social and political government." On a band around the sheaf these canny masters emblazoned the motto "*Aliis Exterendum,*" "It must be threshed by others," as though the facts threshed aren't conditioned by the gathering and sheaving. But even the titles of the journal's first papers gave them away: "Social and moral statistics of criminal offenders," "Vicious extent and heavy expense of advertisements in England," "On the accumulation of capital by the different classes of society." They kept the wheatsheaf but dropped the motto in 1857.[2]

Occasionally even today a neutral gray waistcoat is donned against suggestions that the work of science and technology is value-freighted. But, just as in Hooke's day, and in Whewell and Babbage's, no major creative activities in society—especially those that are pragmatically or symbolically powerful—should be allowed to claim valuative or moral immunity.

Jacob Bronowski's distinction is key: "Those who think that science is ethically neutral confuse the findings of science, which are, with the activity of science, which is not" (1965, p. 63). Research, once accomplished, must be considered in the long run ambi-potent, usable for either good or evil. The anticholinesterase chemicals developed as nerve gases between the World Wars later turned out to be elegant research weapons in the protein biochemistry revolution, and botanical research on how trees drop their leaves in autumn led to development of the military defoliant Agent Orange.[3] In the short run, of course, facts and know-how can be kept secret, or applied under close control, but they are likely to be revealed or discovered independently elsewhere, eventually.

But: We must vigorously resist any notion that researchers are helpless to make choices among envisionable future lines of research and development, or among possible conditions of pursuit. Although it may not be

2. For history of the Statistical Society, see Mouat, 1885, and Cullen, 1975.

3. Arthur Galston, 1972, chronicles the defoliant history.

useful to regard already published knowledge as having any particular moral or ethical cast, surely it is wrong to view not-yet-accomplished research, which cannot be undertaken without commitment of will and resources, as being anything other than value-laden. And, because timing and pacing always are important, it would be naïve not to recognize that new knowledge may at the moment of its emergence have maleficent or beneficent potency that demands attention. Questions of practical ethics always lie in *what to do next.*

Technical activity must be considered value-laden in two senses: technical people's social values and value perceptions affect their research and service; and that work, in turn, affects the value-situation of others in the public.

Thus I consider to be value-laden: the undertaking and supporting of research (for example, committal of funds to research on acquired immune deficiency syndrome (AIDS), or other disease); the choosing of conditions of experimentation (diplomatic auspices of hurricane-seeding experiments in the South Pacific Basin); the marshaling of science to analyze, assess, and help decide on socially important problems (agricultural policymaking); the investigating of people's values, using social-scientific methods (studying jail guards' attitudes); the applying of technology and medicine to practical problems (treatment of breast cancer); and the incorporating of scientific knowledge into the fabric of social philosophies and policies (taking the findings of child psychology and of income-maintenance economics into account when revising child welfare programs).

Technical experts make crucial decisions for, and in the name of, the public. Although geology students usually don't view themselves as moving into a value-charged realm—what could be less social than rocks, after all?—as their careers progress geologists find themselves making seismicity assessments for hydroelectric or nuclear power plant siting decisions, advising on beach protection, municipal building codes, transnational water resources, seabed mining, and strategic minerals supply, and leading projects on causes of acidification of lakes and on underground disposal of radioactive waste. Teams of automotive engineers design cars and sell the designs through corporate management to the public. Pharmaceutical experts develop, test, and push drugs toward the market. Nuclear power plant designers weight the ratio of instant to delayed (cancer) death risks they design into reactors. Nuclear managers decide, in cleaning up after an accident, between exposing a few workers to radiation for relatively long times and exposing more workers for shorter times. Much research by social scientists—on the effect of school busing on educational achievement, on the effect of incarceration on criminal recidivism, on the influence of wage incentives on acceptance of occupational hazard—is so integral to policymaking that analysis can hardly be distinguished from advocacy. And of course some scientists and physicians themselves become high official decisionmakers in industry, labor, and government.

Experts' overt value stances can be argued about. Much harder to dig out and deal with are inarticulate premises. Genetic counselors' counsel is bound to be tempered by their attitude toward contraception and abortion. Marine ecologists' advice on the dumping of wastes into the ocean hinges on whether they think of the oceanic environment as being fragile or resilient, and on whether they prefer a pristine ocean to a "working" ocean. In psychotherapy, as Anne Seiden has argued, "the assumption that dependency, masochism, and passivity are normal for women and the tendency to treat assertiveness and aggression differently for women than for men" leads to "different standards of health for women and men"; therapists' practice thus is conditioned by their intuited, schooled, and inferred interpretation of gender (1976, p. 1116).

Later chapters will address such questions that arise on this account as: How should advisors and advisory committees, in their procedures and reports, deal with their factual biases and value preferences? (As with bias in a textile, "bias" simply means inclination, and isn't necessarily pejorative.) Since most of the work individual scientists do is guided by personal motives and morals rather than by grand ethical schemes, and since each researcher contributes only small increments to the overall technical enterprise, how should individual scientists' actions be oriented to "society's" values? How should scientific research freedoms be balanced against societal constraints?

Serious trouble arises when the distinction between facts and values is blurred or not recognized, or when disputants engage in mislabeling. Nothing has illustrated this more dramatically than the congressional abortion battle of 1981. Senate Bill 158 was introduced which would extend to unborn fetuses the rights of due process guaranteed by the Constitution's Fourteenth Amendment, circumventing Supreme Court decisions preserving women's right to abortion. In hearings, five medical researchers and physicians were drafted into testifying that a human being is formed at the moment sperm fuses with egg, and that this is a "scientific fact." Professor Jérôme Lejeune of the Medical College of Paris asserted, "To accept the fact that after fertilization has taken place a new human has come into being is no longer a matter of taste or of opinion. The human nature of the human being from conception to old age is not a metaphysical contention, it is plain experimental evidence." Boston physician Micheline Mathews-Roth insisted that "one is being scientifically accurate if one says that an individual human life begins at fertilization or conception." Entering the fray, Yale geneticist Leon Rosenberg protested that "the notion embodied in the phrase 'actual human life' is not a scientific one, but rather a philosophic and religious one." He quoted geneticist Joshua Lederberg: "'Modern man knows too much to pretend that life is merely the beating of the heart or the tide of breathing. Nevertheless he would like to ask biology to draw an absolute line that might relieve his confusion. The plea is in vain. There is no single, simple answer to 'When does life begin?'.'" Rosenberg emphasized, "I have no quarrel with anyone's ideas on this matter, so long as it is clearly understood that they are personal beliefs . . . and not scientific truths" (U.S. Senate, Subcommittee on Separation of Powers, 1981). The transgression was important enough to move the National Academy of Sciences to pass one of its rare resolutions (April 28, 1981):

> It is the view of the National Academy of Sciences that the statement in Chapter 101, Section 1, of U.S. Senate Bill S158, 1981, cannot stand up to the scrutiny of science. This section reads "The Congress finds that present day scientific evidence indicates a significant likelihood that actual human life exists from conception." This statement purports to derive its conclusions from science, but it deals with a question to which science can provide no answer. The proposal in S158 that the term "person" shall include "all human life" has no basis within our scientific understanding. Defining the time at which the developing embryo becomes a "person" must remain a matter of moral or religious values.

FACT/VALUE INTERPLAY, FROM SOCIAL DARWINISM THROUGH WILSONIAN SOCIOBIOLOGY

Some of the most extreme exploitations of science in moral debate occurred around Social Darwinism late in the nineteenth century. During this clamor every major social movement struggled to accommodate the unsettling revelations of Wallace and Darwin (or popular interpretations of their theories), while at the same time defending its own ingrained views of class, wealth, race, sex, progress, and justice. The "moral economy of nature" and the "vital order of society" were analyzed in terms of specialization of function and social equilibrium. Spencer justified laissez-faire capitalism by equating economic competition with natural selection, and he drew upon a revised Malthusianism to explain why poverty was unavoidable. Engels fought the trend by pointing out how economic activity intervenes in selection, and how, as Darwin himself had recognized, cooperative behavior can enhance survivability. Comte represented himself as siring sociology out of the biological sciences. And later, on an openly Darwinian theme, Galton founded the eugenics movement. Reference to evolutionary science often was simply sham scientism dragged in for justification, but many eminent scholars' excesses stemmed from sincere internal struggles.[4]

4. Greta Jones, 1980, is a superb analysis of Social Darwinism. Hofstadter, [1944] 1959, is the classic American account. Kelly, 1981, describes the partial Darwinization of Marxism circa 1870.

Many of the utopian visions that have come along since then have yearned vaguely for a society founded on somehow-scientific principles. In medicine, for instance, Frederick T. Gates, the Baptist minister and chairman of the board of The Rockefeller Institute for Medical Research, on that institution's tenth anniversary in 1911 promised (Corner, 1964, p. 4):

As medical research goes on, it will find out and promulgate, as an unforeseen byproduct of its work, new moral laws and new social laws—new definitions of what is right and wrong in our relations with each other. Medical research will educate the human conscience in new directions and point out new duties. It will make us sensitive to new moral distinctions.

Engineers similarly have reached for moral "rationality." In his presidential address to the American Institute of Electrical Engineers in 1919 Comfort Adams wondered (1919, p. 792):

Are there no laws in this other realm of human relations which are just as inexorable as the physical laws with which we are so familiar? Is there no law of compensation which is the counterpart of our law of conservation of energy?

And in the heady 1910 of the University of Chicago, Albion Small foresaw social scientists serving as "sailing masters" for the nation (1910, p. 242):

The most reliable criteria of human values which science can propose would be the consensus of councils of scientists representing the largest possible variety of human interests, and co-operating to reduce their special judgments to a scale which would render their due to each of the interests in the total calculation.

One stream of latter-day Social Darwinism began with the great English biologist T. H. Huxley at the turn of the century, was modified two wars later by his grandson Julian, and was elaborated upon at mid-century by C. H. Waddington. In *The Ethical Animal* Waddington hypothesized that "the function of ethical beliefs is to mediate human evolution, and that evolution exhibits some recognizable direction of progress" (1961, p. 59). Acerbically he noted, "The horrifying effects of social actions based on excessive beliefs of an allegedly ethical character, as they are exhibited in the wars and persecutions in the name of religion, politics, nationalism, racism and various other idealisms, is sufficient evidence that the human condition might well be improved" (p. 203), leading him to conclude that "the major ethical problems of today in the context of individual-to-individual behaviour would, I think, according to our criteria, have to be sought in those types of attitude and activity which facilitate or hinder the development of a healthy authority structure" (p. 205). But he didn't specify how the search should proceed, or what authority structures were to be considered healthy (T. H. Huxley and Julian Huxley, 1947; Waddington, 1961).

From time to time social scientists have tried to derive cultural guidance from what they see as historical "moral evolution," for example from sanctions observed in primitive cultures and in utopian communities, and from analogues to human morality, such as altruism and sexual fidelity, perceived in the behavior of lower animals. I don't hold much hope for most of these efforts. The analogical constructions seem quite shaky. Despite promises, these studies have not generated any specific guidance on real-world problems.[5]

The latest phase in this line of propositions began in 1975 when Harvard biologist Edward O. Wilson published *Sociobiology: The New Synthesis*. The book's central theoretical query was: "How can altruism, which by definition reduces personal fitness, possibly evolve by natural selection?" Wilson outlined how he would answer (1975a, p. 3):

> The hypothalamic-limbic complex of a highly social species, such as man, "knows," or more precisely it has been programmed to perform as if it knows, that its underlying genes will be proliferated maximally only if it orchestrates behavioral responses that bring into play an efficient mixture of personal survival, reproduction, and altruism. Consequently, the centers of the complex tax the conscious mind with ambivalences whenever the organisms encounter stressful situations. Love joins hate; aggression, fear; expansiveness, withdrawal; and so on—in blends designed not to promote the happiness and survival of the individual, but to favor the maximum transmission of the controlling genes.

Building upon his renowned lifelong research on insect societies and comparative biology, for twenty-six chapters Wilson conducted an enchanting tour, albeit with disturbing overtones, of African ant and termite nests, Scottish barnacle colonies, Montana sagebrush grouse leks, and Serengeti wildebeest herds, up through equatorial African chimpanzee societies. Then came the concluding, overreaching Chapter 27, "Man: From Sociobiology to Sociology," which carried a section on ethics that began: "Scientists and humanists should consider together the possibility that the time has come for ethics to be removed temporarily from the hands of the philosophers and biologicized."

Three almost baiting propositions (among the most egregious) from Wilson, from *Sociobiology* and related publications, will illustrate why he drew criticism. More-contentious theses hardly can be imagined. On division of

5. Flew, 1967, proposes a search for "evolutionary ethics." Donald Campbell, 1975, makes provocative suggestions. Stent, 1980, collects representative wishful thinking. Sperry, 1983, speculates, unproductively to my reading, on the mind–brain issue in relation to values.

labor between genders (1975b, p. 48):

> In hunter-gatherer societies, men hunt and women stay at home. This strong bias persists in most agricultural and industrial societies and, on that ground alone, appears to have a genetic origin.

On religious indoctrinability (1975a, p. 561):

> The enduring paradox of religion is that so much of its substance is demonstrably false, yet it remains a driving force in all societies. Men would rather believe than know, have the void as purpose, as Nietzsche said, than be void of purpose. . . . Human beings are absurdly easy to indoctrinate—they *seek* it.

On rights (1976, p. 189):

> To the extent that the biological interpretation noted here proves correct, men have rights that are innate, rooted in the ineradicable drives for survival and self-esteem, and these rights do not require the validation of ad hoc theoretical constructions produced by society.

Wilson came under attack in part because he made some strong assertions that he didn't (and I think couldn't) defend, in part because sporadically he lapsed into fuzzy language, and in part because he was presumed guilty of having Spencerian tendencies. Some critics then made sociobiology into a proxy debate over sexism, economic justice, and other perennial issues.[6]

Most reviewers have found Wilson's ethological analyses and interdisciplinary coverage stimulating, though not novel, but have objected strenuously to his more extreme speculations. On that twenty-seventh-chapter leap, Wilson's Harvard colleague Stephen Jay Gould remarked (1977, p. 252):

> We who have criticized this last chapter have been accused of denying altogether the relevance of biology to human behavior, of reviving an ancient superstition by placing ourselves outside the rest of "the creation." Are we pure "nurturists?" Do we permit a political vision of human perfectibility to blind us to evident constraints imposed by our biological nature? The answer to both statements is no. The issue is not universal biology vs. human uniqueness, but biological potentiality vs. biological determinism.

---

6. Key publications on this phase of sociobiology include Wilson, 1975a, 1975b, 1978; Caplan, 1978; Gregory, Silvers, and Sutch, 1978; Barash, 1979; Chagnon and Irons, 1979; Freedman, 1979; Bock, 1980; Hunt, 1980; Montagu, 1980; Lumsden and Wilson, 1981; Leeds and Dusek, 1981–1982; and Wiegele, 1982.

Gould then asked (1977, p. 257):

> Why imagine that specific genes for aggression, dominance, or spite
> have any importance when we know that the brain's enormous flexi-
> bility permits us to be aggressive or peaceful, dominant or submissive,
> spiteful or generous? Violence, sexism, and general nastiness *are* bio-
> logical since they represent one subset of a possible range of behav-
> iors. But peacefulness, equality, and kindness are just as biological—
> and we may see their influence increase if we can create social struc-
> tures that permit them to flourish.

Especially harsh criticism came from a Boston Sociobiology Study
Group, an affiliation of the leftist organization Science for the People.
Acrimoniously the Group attacked *Sociobiology*, "the manifesto of a new,
more complex, version of biological determinism," as drawing unfounded
analogies between nonhuman and human societies, as engaging in "speculative reconstructions of human prehistory," as overemphasizing the
genetic bases of behavior, and as making untestable assumptions about
selection-adaptive drives in behavior. Wilson countercharged them with
practicing "academic vigilantism" (Allen et al., 1975; Wilson 1976; Wade,
1976).

My view is that although Wilson pulled together some fascinating themes
in a provocative way, he overreached badly in his social speculations. To
Wilson's claim that a "neurologically based learning rule" makes humans
"absurdly easy to indoctrinate," literary critic Stuart Hampshire admonished, "Vast obscurities are concealed in that phrase 'neurologically
based'" (1978, p. 65), and Stephen Gould shrugged, "I can only say that
my own experience does not correspond with Wilson's" (1977, p. 254).
Philosopher Ruth Mattern expressed my own reservations in saying, "A
form of sociobiology attenuated enough to be plausible seems to be too
weak to take ethics out of the hands of philosophers" (1978, p. 470).

Its unfortunate features aside, the sociobiology controversy has stimulated vigorous interdisciplinary discussion, and it continues to reward following as an example of is/ought exploration.

MEANING OF "VALUES" AND "VALUE"

"Values" and "value" have been given so many connotations by the public
and by philosophers, theologians, psychologists, economists, and other
specialists that I cannot hope to refine them into neat definitions. What I
will try to do is draw some commonsense perspective, illustrate how values
are manifested, and show how values enter into analysis and decisionmaking.

In ordinary usage, *values* are taken to be abstract aspirations: freedom
of speech, cohesive family life, national security. Such goals may be neither

perfectly definable nor attainable, but, as with the Constitution of the United States, they can serve as ideals. It is in this spirit that the World Health Organization constitution declares that all people have a right to the "highest possible level of health," which it defines as "a state of complete physical, mental, and social well-being, not merely the absence of disease or infirmity."

Values can, in narrower usage, be taken as potentially attainable states of affairs, as objectives: assurance of infant survival, eradication of leprosy, achievement of energy independence, maintenance of forests for future generations. Values can govern means as well: protection of entrepreneurial access to seabed minerals, requirement that behavioral experiments not be carried out unless the subjects freely grant informed consent.

Value I take to be ascribed worth, as reflected in social preferences and transactions: market value of zinc, information value of a blueprint, political value of a senator's endorsement, social value of literacy, aesthetic value of a cityscape, symbolic value of a new medical clinic.

Value-laden (*-freighted*, *-charged*, *-oriented*) then connotes that an analysis, decision, or action is influenced by personal or institutional proclivities and prejudices, and that the analysis, decision, or action may affect people's value situation—their opportunities, status, wealth, happiness, or aspirations.

In 1752 David Hume made clear how valuation regresses to deep "sentiments":[7]

> Ask a man *why he uses exercise*; he will answer, *because he desires to keep his health*. If you then inquire *why he desires health*, he will readily reply, *because sickness is painful*. If you push your inquiries further and desire a reason *why he hates pain*, it is impossible he can ever give any. This is an ultimate end, and is never referred to any other object.
>
> Perhaps to your second question, *why he desires health*, he may also reply that *it is necessary for the exercise of his calling*. If you ask *why he is anxious on that head*, he will answer, *because he desires to get money*. If you demand, *Why? It is the instrument of pleasure*, says he. . . . Something must be desirable on its own account, and because of its immediate accord or agreement with human sentiment and affection.

Three fundamental questions have pervaded value-laden decisions and actions throughout history. How should choices be made when options cannot all be pursued or when they conflict? (If people are differently vulnerable to health hazards, and workplaces can never be perfectly risk-free, how should equal health protection be reconciled with equal employment opportunity?) How should collective societal goods be pursued with least erosion of the rights and goods of affected individuals? (What degree of vaccination efficacy for a population should be judged to outweigh side-

---

7. First appendix to *Inquiry Concerning the Principles of Morals*, 1752.

effect risks to individuals?) And how should specific guidance on particular real actions be derived from abstract high precepts? (What does "right to personal privacy" mean in our present world of electronic financial transactions, international campaigns against terrorism and drug smuggling, and institutionalized health recordkeeping?)

People's strivings toward what they value—home and neighborhood lifestyles, opportunities for their children, assistance to citizens of less fortunate countries—hardly can be expressed precisely. Within any group, preferences will differ, and values will change over time. It is especially hard to resolve grand goals (national security) into the prosaic objectives (missile deployment plans, titanium stockpile policies, computer export restrictions), themselves value-laden, required for achieving those goals.

Public opinion polls can to some extent reveal value concerns. I remain a skeptic, and believe that polls that don't force respondents to choose among real options and confront trade-offs are worthless. Actions speak louder than answers to surveys. To me, actual manifestations of valuation—in political actions, budgets, laws, treaties, regulatory policies, military strategies; in court, corporation, and labor union decisions; in consumer purchasing behavior; in medical preferences; and in wage differentials and insurance schemes—are much more telling than casually expressed "opinions" are.

Throughout this book I will explore how value concerns and preferences can be elicited and analyzed, and how technical assessments, decisions, and activities can be linked to societal ethics and goals. Chapter Eight will consolidate and review these matters at length.

INFLUENCE OF SCIENCE AND TECHNOLOGY ON SOCIAL PHILOSOPHIES AND CHOICES

Although they overlap and are not a formal taxonomy, the following modes of influence can be distinguished.

Science deeply informs our cultural outlook. Science has transmuted quite a few major cultural myths; negated many superstitions; left us living in a "dis-enchanted" world; imparted substance to a host of miasmas, humours, auras, scourges, and vital forces; recast the mind–body, nature–nurture, and other classic mysteries; and conspicuously revealed the hand of Man where none was seen before but Fortune's. Science reveals fundamentals about the occurrence and causes of mortality, genetic inheritance, and material wealth; gives us insights into where we have come from, and into our place in the universe; enables us to understand how we perceive what we see and mean what we say; and not only describes particular cultures, but helps us elaborate the very notions of "culture" and "society."

Scientific and technological advance can create options for public consideration.
Choices from intimately personal to Malthusianly global have been opened
up by the invention of the condom, diaphragm, spermacidal foam, Pill,
and IUD, while other choices remain distant because of the practical
unavailability of a male Pill, reversible vas deferens valve, and other contraceptive options. In addition to such options for *doing*, technology can
create options for *knowing* (and then perhaps doing): until the recent
development of amniocentesis and related techniques, never had it been
possible to know with any certainty the gender, genetics, or pathology of
a fetus in utero and thus be presented with informed choices over carrying
to term, or seeking pre- or postnatal therapeutics, or terminating the
pregnancy.

*Technology can strongly alter the relative attractiveness of competing social
alternatives, and can induce value changes.* The increasing practicality of
solar energy and conservation methods will have implications, in the long
run, for issues ranging from insulation installers' pulmonary health to diplomatic relations with Persian Gulf sheikdoms. Advance toward such
dreams as rooftop solar electric cells, or vaccines against venereal diseases,
depends not just on more efficient manufacture or adaptation, but on fundamental scientific discovery. Possibility changes tend to induce value
changes, although they don't necessarily do so: the development of surgical anesthesias in the nineteenth century led to revision of the risks and
costs patients were willing to bear in the surgical "calculus of suffering"
(Pernick, 1985).

*Science can identify and analyze consequences of choices and events, and help
raise issues to public attention.* Science describes causal conditionals. Social
attitudes and actions are altered by the knowledge that masturbation does
not, contrary to earlier dogma, lead to madness; that pellagra, far from
being an inherited inferiority, is the result of a dietary niacin deficiency
that can easily be remedied; that some forms of schizophrenia can be
attributed not to a mystically evil soul but to treatable physiological anomalies; and that soft, fluffy, seemingly harmless textile dust causes brown
lung disease. Such knowledge allows causes to be assigned, opportunities
and liabilities identified, and ethical issues altered.

*Science can anticipate and analyze perturbations in society itself, including
impacts of technological change.* It can estimate the effects of entering alternative energy futures, of building a high-speed train system between Los
Angeles and San Francisco, of mandating kindergarten attendance. It can
project demographic changes, the relation of future populations to the
resources they will have available, and the likely future interactions
between people and technologies.

*The social sciences can observe and analyze expressed and implicit social values
and valuation processes.* Reports on social attitudes and practices, such as
surveys of sexual habits, can prime the way for changes in the way people
evaluate their own attitudes. Although social scientists from Max Weber

through Clyde Kluckhohn to the present have hoped to become able to analyze values and valuation psychologies, their methods are just now gaining enough acuity to warrant practical use. Social scientists are making progress in constructing value typologies, in surveying voter, medical client, and consumer attitudes, and in analyzing the preferences implicit in economic and legal actions. All of this information can be brought to bear on social decisions.

TRACTATUS

The argument so far has been intended to establish these fundamentals.

- Social values cannot be derived from science qua science alone.
- Although scientific knowledge, once attained, may be considered ambi-potent for good or evil, the work of pursuing new science and developing technologies, which requires commitment of will and resources to undertake, is by no means value-neutral.
- Besides, at the moment of its emergence new knowledge may well have maleficent or beneficent potency that demands attention.
- Technical activity must be considered value-laden in two senses: technical people's social values and value perceptions affect their research and service; and that work, in turn, affects the value-situation of others in the public.
- Technical experts make crucial decisions for, and in the name of, the public.
- Science and technology affect our philosophies and choices by: deeply informing our cultural outlook; creating options for public consideration; altering the relative attractiveness of competing alternatives, and inducing value changes; identifying consequences of choices and events, and helping raise issues to public attention; anticipating and analyzing perturbations in society itself, including impacts of technological change; and observing and analyzing expressed and implicit social values and valuation processes.

TECHNICAL PROGRESS AS DIRECTED TRAGEDY

Because it pervades this book's outlook, I must now introduce my view of technical progress as tragedy. Not mere sadness or misfortune, but tragedy in a high sense. And not fatalistic tragedy, but, in what has become a characteristic of modern society, deliberately directed tragedy.

"The myths warn us that the wresting and exploitation of knowledge are perilous acts, but that man must and will know, and once knowing, will not forget," David Landes has reminded us (1969, p. 555).

Adam and Eve lost Paradise for having eaten the fruit of the tree of knowledge; but they retained the knowledge. Prometheus was punished, and indeed all of mankind, for Zeus sent Pandora with her box of evils to compensate the advantages of fire; but Zeus never took back the fire. Daedalus lost his son, but he was the founder of a school of sculptors and craftsmen and passed much of his cunning on to posterity.

Man—*bestia cupidissima rerum novarum*, the "species cupidinous of new things"—must and will know. That ambitiousness has long been embodied in our Western tragic sense of ourselves. Alfred North Whitehead provocatively recast it (1925, p. 14):

The pilgrim fathers of the scientific imagination as it exists today are the great tragedians of ancient Athens, Aeschylus, Sophocles, Euripides. Their vision of fate, remorseless and indifferent, urging a tragic incident to its inevitable issue, is the vision possessed by science. Fate in Greek Tragedy becomes the order of nature in modern thought.

Crucially, Whitehead continued: "Let me remind you that the essence of dramatic tragedy is not unhappiness. It resides in the solemnity of the remorseless working of things." And I would add the emphasis, " . . . especially as human agency intervenes."

For present purposes, I take tragedy to mean the deliberate confrontation of deeply important but nearly irresolvable life issues. Tragedy begins in our knowing of causalities, in our intervening in particular causes, and in our technical enlargement of interventional possibilities.

Robert Oppenheimer's confesso—"In some sort of crude sense which no vulgarity, no humor, no overstatement can quite extinguish, the physicists have known sin"—sounds only to be a starkly lame *mea culpa*, unless one realizes that the ages-old service of science and technology to warmaking was too well known to Oppenheimer and his colleagues (1948, p. 66). No; surely the emphasis was on the verb: "Physicists have *known* sin." And not bad-boy sin, but Original sin. I think it not unlikely that the father of the A-bomb would approve of my transmutation: *scientists have known tragedy*. It is in that knowing that many of the issues of this book reside.

Nowhere has this "remorseless working of things" been more profoundly evident than in the development of nuclear weapons. In a *Discovery* editorial in September 1939 C. P. Snow, noting that the idea of explosive chain reaction had become accepted among leading physicists, grimly predicted that a project to make an atomic bomb would "certainly be carried out somewhere in the world." The Manhattan Project, of course, went forward, as did its Japanese and German counterparts. Looking back on the 1945 decision to drop the Hiroshima weapon, Robert Wilson recalled (1970, p. 32):

Things and events were happening on a scale of weeks: the death of Roosevelt, the fall of Germany, the 100-ton TNT test of May 7, the bomb test of July 16, each seemed to follow on the heels of the other. A person cannot react that fast. Then too, there was an absolutely Faustian fascination about whether the bomb would really work.

Similarly Norbert Wiener's melancholia of 1948 over the development of cybernetics, although it strikes me as giving in too easily (1948, p. 28):

Those of us who have contributed to the new science of cybernetics stand in a moral position which is, to say the least, not very comfortable. We have contributed to the initiation of a new science which embraces technical developments with great possibilities for good and for evil. We can only hand it over into the world that exists about us, and this is the world of Belsen and Hiroshima. We do not even have the choice of suppressing these new technical developments. They belong to the age. . . .

Once arcane knowledge is generated somewhere, its transmission, or independent reconstruction, is almost, though not absolutely, inevitable. As Dürrenmatt had Möbius say in *The Physicists*, "What has once been thought can't be unthought." After basic nuclear information was released after World War II, the question of proliferation became not *whether* but *when* and *under what circumstances* other countries would pursue nuclear options—and, in Whitehead's phrase, "tragic incident moved to inevitable issue."

Three fundamental tragic motifs can be recognized.

First, by describing flatly the way things are, science raises tragic awarenesses: that certain things are happening, others may happen, others will happen, others cannot happen; that events are determined by causes; that causes may reflect willful human agency and decision.

Even as scientists achieve long-sought humanitarian breakthroughs, the timeless lament of Ecclesiastes (1:18) resonates: "He that increaseth knowledge increaseth sorrow" ("sorrow," or "mental anguish," is a standard translation of the Hebrew, *mak'ôbâh*).

In absence of knowledge, people may resign themselves to Fate, to "blind chance" (mongoloid birth just happens). With knowledge (the chance of Down's syndrome increases sharply with maternal age above thirty-five), intentionality issues arise; chances still have to be taken, but odds may be altered or stakes adjusted deliberately. Tragedy lies not in resigned fatalism, but in considered confrontation of the near-irreconcilables (wanting to have a child, but wanting to pursue other early-life goals first, but of course wishing that the child not be born infirm).

Biomedical science is forcing us to confront basic facts about differences among people, such as differences in allergic vulnerability, color perception, reflex quickness, and lower-back resiliency, that can bring occupational health protection squarely into conflict with equal employment opportunity. Differences we have pretended don't exist will have to be recognized.

Second, in their inventions and in the systems they weave our lives into, technology and medicine confront us with tragic choices among life-expensive options whose consequences we have at least some foreknowledge of.[8] In a trend that can only lead, eventually, to anguishing decisions, improvements in neonatal care, coupled with a desire to save all infants no matter what, are preserving ever-more-premature babies (down to 500 grams, or a little over a pound, birthweight, now), at ever-increasing costs, with higher incidences of permanent medical deficiencies.[9] As society invests in life-extending medical technologies, sorts out endangered-species protection priorities, and debates the future of fourth-world nations, exceedingly traumatic choices will have to be confronted—not just made (we do that already, often by defaulting), but *confronted.* Weighed. Debated. Faced.

The answers do not reside in knowledge by itself. Again we hear a classic voice, Tiresias moaning when Oedipus commanded him to consult "bird-flight or any art of divination" to guide Thebes from the plague: "How dreadful knowledge of the truth can be when there is no help in truth."

Third, technological advance challenges society with tragic commitments to consequences. Warning that commercial nuclear power comes as a "Faustian bargain," in 1972 Alvin Weinberg, the director of Oak Ridge National Laboratory, said "the price that we demand of society for this magical energy source is both a vigilance and a longevity of our social institutions that we are quite unaccustomed to" (1972a, p. 33). Regardless of future decisions about nuclear weapons or nuclear power, the high-level radioactive waste already accumulated in many countries will demand curatorship for thousands of years; the stuff will not go away. The great system of dikes that creates and protects one-third of the Netherlands requires similar massive perpetual commitment. Now that we have eradicated smallpox worldwide as a clinical entity, we shall forever have to monitor our increasingly unvaccinated populace, to watch against recrudescence of the virus from who-knows-what lurking source.

These tragic awarenesses, choices, and commitments caused or mediated by technical ventures may leave us happy or not. But of their solemnity there can be no question.

Think for a moment about our progress in dealing with public health risks, in which all of these themes are so evident. Health in the industrial West surely is, in general, more robust than ever before. Many of the most dangerous infectious diseases, such as tuberculosis, diphtheria, smallpox, cholera, typhus, and polio, have been conquered, and progress has been

8. Calabresi and Bobbitt, 1978, ably essays on some life-and-death decisions, but despite its title only lightly addresses their "tragic" nature.

9. Phibbs, Williams, and Phibbs, 1981; and Murray and Caplan, 1985.

made against many others. Scurvy, pellagra, iron deficiency anemia, and other nutritional diseases have been mastered. Many illnesses that have not yet been eliminated, such as diabetes, have at least been brought under control. Exposure to mercury, lead, arsenic, chromium, and other heavy-metal poisons has been substantially reduced, as has exposure to asbestos, halocarbon solvents, and many other chemicals. Through prediction and protection much damage from hurricanes, floods, and earthquakes has been mitigated. All over the world infant mortality continues to decrease, and life expectancy to increase. More people are living longer, healthier, more vigorous lives.

Nonetheless, almost ruefully, we have progressed to an inherently discomfiting state, a state in which we must expect to remain from now on. Why inherently discomfiting? Because steadily we have broadened our apprehensions to include not only natural catastrophes, infectious diseases, everyday mechanical accidents, and acute poisons, but also large-scale technological accidents, chronic low-level hazards from chemicals, radiation, and noise, and lifestyle vices, such as addictions to tobacco, alcohol, barbiturates, narcotics, caffeine, and rich foods. To our struggle against cancer and other classical illnesses we have added concern about reproductive, genetic, immunological, behavioral, and other debilitations. And of course we continue to create new hazards, to identify risks that have existed without being recognized, and to resolve to reduce previously tolerated risks.

Now, about many hazards we know enough, scientifically, to "worry," but not enough to know *how much* to worry—or how much protective action to invest. Knowledge has grown enormously, and we even have the luxury of going around searching for possible trouble. But many scientific disciplines are still in their adolescence and are unable to evaluate risks precisely. We can detect minuscule traces of manmade pesticides in mothers' milk all over the world, which is vaguely disturbing; but, with rare exceptions, we don't have a clue as to whether the chemicals exert any effect on mother or infant (Jensen, 1983; Wolff, 1983). Toxicology, epidemiology, and medicine have taken us out to their borders, but it's unruly territory.

At the same time that we are learning more, we are heightening our societal aspirations beyond all previous limits. We intend to help all infants get a vigorous start in life. And we strive to afford first-rate, broadly defined health protection to all citizens and noncitizens, even immigrant aliens, through an enormous range of risks, throughout their lives. No civilization ever before has had these ambitions.

The crux: In our knowing so much more, though imperfectly, and aspiring to so much more, we have passed beyond the sheltering blissfulness of ignorance and risk-enduring resignation. This has given rise to considerable social apprehensiveness, which is affecting both the outlook of individuals and the functioning of institutions.

Similar phase-changes have occurred historically when people became aware of specific causes of disease and deformity, as when it became clear that moral turpitude alone was not the cause of syphilis, and when societal aspirations, such as commitment to worker protection, rose. We are going through both kinds of change at the same time.

Risks and aspirations will continue to evolve. In his 1803 revised *Essay on Population* Thomas Malthus observed of Jenner's new vaccine, "I have not the slightest doubt that if the introduction of cowpox should extirpate the smallpox, we shall find . . . increased mortality of some other disease" (1803, p. 522). Malthus was right. As any risk is reduced, others inevitably increase in the mortality and morbidity tables—though perhaps setting in at later ages. Risks in the industrial West are evolving now as rural and agrarian risks are succeeded by urban, industrial, and medical-care risks. A similar progression is occurring, displaced in time, in less developed countries: cholera and fatal infant diarrhea are being succeeded by cancer, heart disease, and drug side effects. I can't imagine that our societal aspirations won't expand even further, both for the industrial West and for the rest of the world.

"We are in for a sequentiality of improbable possibles," *Finnegans Wake*'s Shem knew to expect. Some possibles are, of course, more predictable than others. Recent years' debates over such matters as food additives, contraceptives, pesticides, and energy sources have brought broad public recognition that nothing can be risk-free, that there are no rewards without risks, and that risktaking for benefit is the essence of human striving.

Our analyses and decisions are making us face risks ever more explicitly and comparatively. The Occupational Safety and Health Administration's 1983 standard for worker exposure to airborne inorganic arsenic was a striking example. In tightening the standard from 500 to 10 micrograms arsenic per cubic meter of air, OSHA concluded "that inorganic arsenic is a carcinogen, that no safe level of exposure can be demonstrated, and that 10 micrograms per cubic meter is the lowest possible level to which employee exposure could be controlled." Further (U.S. Occupational Safety and Health Administration, 1983, p. 1867):

> The level of risk from working a lifetime of exposure at 10 micrograms per cubic meter is estimated at approximately 8 excess lung cancer deaths per 1000 employees. OSHA believes that this level of risk does not appear to be insignificant. It is below risk levels in high risk occupations but it is above risk levels in occupations with average levels of risk.

Surely this rationale is commendable (regardless of whether the particular numbers adopted survive current dispute), but the explicitness is unsettling.

And so, now, on into the book. I must emphasize that even while I stress the essentially tragic nature of our awarenesses, decisions, and commitments, I believe that we are in a great many ways better off than ever before. Just as surely as science and technology are part of the problem, they will be part of the solution. And I would affirm that tragic confrontations are an essence of humanness.

The difference from the past is that—discomfiting though it may be— the conjunction of greatly improved technical abilities with heightened societal aspirations now puts us in a position to *direct* aspects of tragic progress. Experts perform center stage and in the wings. All of us speak from the citizens' chorus.

Two

The Changing
Sociotechnical Milieu

That science, technology, and medicine are pervasive and powerful elements of modern life is beyond dispute. But in what senses are they pervasive and powerful? How politicized has technical endeavor become? Does that add up to technocracy? What are the peculiarly modern characteristics of current technologies? What force should we accord the propositions of technological momentum and determinism?

Clearly a great many endeavors that once were intuitive, unsystematized, or a matter of knack have—not necessarily effectively or legitimately—been stretched onto an armature of analytic and decisional rationality. Thus President Kennedy emphasized to managerial audiences that "most of the problems . . . that we now face are technical problems, are administrative problems. They are very sophisticated judgments which do not lend themselves to the great sort of passionate movements which have stirred this country so often in the past" (1962, p. 422). From time to time observers scorn this trend as undue scientization or technicization. As I will describe shortly, many core aspects of life have been medicalized. For all such issues the question is what constitutes overreaching in pursuit of rationality. Certainly little in life is considered out-of-analytic-bounds anymore. Even sexuality has become scientized: as Lionel Trilling cleverly remarked in a review of Kinsey's *Sexual Behavior in the Human Male*, "*Alma Venus* having once been called to preside protectively over science, the situation is now reversed" (1948, p. 464).

POLITICIZATION AND POLITICKING

Politicization

In earlier times scientific research often was treated as politically neutral, or, at least, as transcending nationalism. In 1779 Benjamin Franklin directed Colonial warships not to hinder Captain Cook's explorations, "because the increase of geographical knowledge facilitates the communication between distant nations and the exchange of useful products and manufactures, extends the arts, and science of other kinds is increased to the benefit of mankind in general" (Wecter et al., 1949, p. 77). During the Anglo–French War Napoleon granted the British chemist Humphry Davy safe conduct to visit the Institut Français.[1] In 1802 the French captured the theodolite of Britain's Great Trigonometrical Survey being shipped to India, but then forwarded the instrument to the Survey with a letter of good wishes (Wilford, 1981, p. 163). Many such examples can be cited.

To this day, scientific correspondence and specimens and instruments are exchanged across national boundaries fairly unimpeded by political constraints. Medical care has always proceeded under a white flag, even through the worst of wars. This ethos of neutrality and universalism persists, in diluted form, in such multilateral undertakings of mutual purpose as the treaty-sanctioned preservation of Antarctica for scientific research, the International Atomic Energy Agency's nuclear fuel safeguards program, and the Mediterranean Sea cleanup sponsored by the United Nations Environment Program. The World Health Organization's final, and ultimately successful, vaccination encirclement of smallpox was allowed to continue, albeit dicily, through the Ethiopian–Somali war of the late 1970s.

However, in recent years, because so much technical work is now supported by the public purse and conducted in institutions, rather than by lone investigators working at personal expense, and because the physical and symbolic import of technical accomplishments has become immense, this ethos has been substantially eroded.

The space race of the 1960s dramatically served notice of the change. The Apollo venture was both an ultimate example of Jacques Barzun's "science as glorious entertainment" (1964) and a projection of massive technical, political power. If I may irreverently paraphrase: "That's one small step for a man, one giant leap for mankind, one hell of a stunt by the ol' U.S.A." Lunar scientists complained that the moonshots were not optimized for scientific exploration, and many astronomers testified that the research return would have been greater if the money had been invested in satellite and telescope projects. Pressed in 1963 by Senator

---

1. DeBeer, 1960, is a superb history of such attitudes in Britain and France 1689–1815.

Stuart Symington on whether the program's budget wasn't growing too fast, Jerome Wiesner, President Kennedy's science advisor, defended "the deep military implications, and very important political significance" of the project, but conceded, "I think if I were being asked whether this much money should be spent purely for scientific reasons, I would say emphatically, 'No.'"[2]

Politicization can lead to ambiguity of purpose. This has occurred, for example, in government-sanctioned technical exchanges. Often such exchanges are conceived as fairly neutral cultural tokens, not very different from artistic and athletic exchanges. So diplomats sign agreements, committees make plans, and, with diplomatic blessing, scholars travel to do research. Sensitive topics are avoided, with, say, yogurt production and botany pursued more avidly than criminology or pollution studies. But then, if intercapital relations become strained, governments may step back in and restrain the exchange. Clumsiness and ineffectiveness in many of the programs initiated in the early 1970s, particularly with the Soviet Union, has led the United States to tighten standards and engage in technical exchanges much more cautiously and with more government control, even in medicine.[3]

Internationally, many impediments are being erected against research. Most nations, led off by the United States, have restricted oceanographic research access within 200-mile coastal zones, making many waters inaccessible; marine research planning now requires long lead times and elaborate diplomatic maneuvering. Laser physics, archaeology, cultural anthropology, and other pursuits that are sensitive because of political or national-security concern, or because of suspicion that the findings won't be published and shared, are becoming hampered.

As applied-science and medical endeavors become big businesses, political and economic considerations may trespass on traditional provider–client relations. The industrialization of medical care has eroded the intimate and private nature of medicine in ways that are familiar and need no expounding here. But the full extent of the change may not be evident, even to doctors. Physician Mark Siegler related this episode (1982, p. 1519):

A patient of mine with mild chronic obstructive pulmonary disease was transferred from the surgical intensive-care unit to a surgical nursing floor two days after an elective cholecystectomy. On the day of transfer, the patient saw a respiratory therapist writing in his medical chart . . . and became concerned about the confidentiality of his

2. U.S. Senate, Committee on Aeronautical and Space Sciences, 1963, p. 34.

3. NAS/NRC, Board on International Scientific Exchange, 1977, reviews U.S.–U.S.S.R. exchanges. U.S. Congress, Congressional Research Service, 1977, discusses many issues of technical exchange, including medical, space, and social-science exchanges.

hospital records. The patient threatened to leave the hospital prematurely unless I could guarantee that the confidentiality of his hospital record would be respected. . . . I was amazed to learn that at least 25 and possibly as many as 100 health professionals and administrative personnel at our university hospital had access to the patient's record and that all of them had a legitimate need, indeed a professional responsibility, to open and use that chart. These persons included 6 attending physicians (the primary physician, the surgeon, the pulmonary consultant, and others); 12 house officers (medical, surgical, intensive-care unit, and "covering" house staff); 20 nursing personnel (on three shifts); 6 respiratory therapists; 3 nutritionists; 2 clinical pharmacists; 15 students (from medicine, nursing, respiratory therapy, and clinical pharmacy); 4 unit secretaries; 4 hospital financial officers; and 4 chart reviewers (utilization review, quality assurance review, tissue review, and insurance auditor).

Probably these involvements accreted quietly, each, as Dr. Siegler acknowledged, under some "legitimate" need. Many other such changes have come along. Together, they add up to an entirely new ethical milieu.

An analogous and intensifying problem is the legal tug-of-war over information relevant to public health decisions. The Procter & Gamble Company, manufacturer of the Rely® tampons that have been associated with toxic shock syndrome, joined in defending academic scientists, to whom it had given unrestricted research grants, against having to release laboratory information deemed preliminary and inconclusive; the courts supported the company's case.[4] Then, wanting to confirm the epidemiological interview findings the government had used in implicating the tampons with causing the disease, Procter & Gamble asked the courts to order the U.S. Centers for Disease Control to reveal the identities of the women interviewed; the agency resisted, on grounds of protecting medical and sexual privacy (Sun, 1984). In part all this has to do with balancing conflicting legal and other interests, and in part it has to do with protecting new scientific and medical information against premature, and possibly abusive, release. The lawyers are everywhere. On occasion government litigants in toxic chemical disputes have refused to divulge health and environmental data, medical investigators have held back on publishing findings until lawsuits reached propitious stages, attorneys have threatened legal action to gain access to scientific symposia, and corporate physicians have been restrained by their company's lawyers from speaking in public.

4. Memorandum opinion and order sur *Rogers and Rogers* v. *Procter & Gamble Co.*, no. 83-0487-C, slip opinion (E.D. Mo., May 4, 1984). Also relevant is *Andrews* v. *Eli Lilly & Co.*, 97 F.R.D. 494, 500 (N.D. Ill., 1983), which denied manufacturers of diethylstilbestrol (DES) access to records of research on the association between pregnant women's ingestion of DES and their daughters' development of vaginal adenocarcinoma.

Politicking

More and more now, technical people engage in organized political action, and not just in promoting their professions' welfare. In its scope and tone this has amounted to substantial overt politicization since mid-century. Notice the wide variety of interests represented in the following examples.

Scientists have protested for and against particular technologies, such as the antiballistic missile system and the supersonic transport aircraft, and they have lobbied on rights to abortion and free speech. In the 1964 presidential election 2,400 psychiatrists indicated in a *Fact* magazine poll that they saw signs of mental instability in candidate Barry Goldwater (this examination-at-a-distance was censured by many psychiatrists and by the American Medical Association). The American College of Obstetricians and Gynecologists has lobbied against federal regulation of sterilization decisions. Physicians for Social Responsibility, defining nuclear war as the ultimate public health threat, campaigns vigorously against the nuclear arms race. The American Medical Association (AMA) operates a Political Action Committee that is one of the most influential, both in funds raised and in political campaign contributions donated. In 1983 the AMA's House of Delegates passed a resolution criticizing the newsmedia's coverage of the dioxin issue, and a resolution opposing the legal insanity defense. Western psychiatrists repeatedly have censured Soviet psychiatric abuses; in reaction, in 1983 the Soviet All-Union Society of Psychiatrists and Neuropathologists withdrew from the World Psychiatric Association. A Council of Professional Associations on Federal Statistics, formed in reaction to the Reagan administration's reduction of budgets for federal statistics-gathering programs, now lobbies in support of the more important of these programs. In 1983, to the expressed shock of physicist and White House science advisor George Keyworth, the Council of the American Physical Society adopted a resolution urging steps in nuclear arms control and reduction.[5] At about the same time a large group of American anthropologists acting "in the name of a discipline whose subject is humanity, and for the sake of humanity" published an advertisement in the *New York Review of Books* (October 13, 1983) protesting that they "unequivocally condemn preparations for war in the name of deterrence."

Naturally, technical people and their organizations also politick on behalf of their own research and professional interests. Scientists avidly campaign for financial support, research access, and advancement of their professions. Some of this lobbying is discipline-oriented, as when academic chemists promote basic inorganic chemical research. Some is problem-oriented, as when medical experts speak for the Committee to Combat Huntington's Chorea. Some amounts to hard-core coalition bargaining, as

5. Keyworth, 1983, and Marshak, 1983, record the resolution, Keyworth's protest, and a response from the president of the American Physical Society.

when space scientists and aerospace industries jointly lobby for orbital spacecraft programs.

No need to belabor the point; the newspapers make it every day. Science and scientists now perform vigorously in the center of most sociopolitical arenas, and they are unlikely ever to be displaced. This raises many questions, which I will discuss throughout the book, about personal roles, institutional roles, technical community roles, and the relations between technical people and their patrons, clients, and the general public.

TECHNOCRACY?

During the 1960s, recognition of the independence and public influence of technical leaders led to their being referred to as a "strategic elite" (Suzanne Keller) of "new Brahmins" (Spencer Klaw) in a "secular priesthood" (Ralph Lapp) of "the republic of science" (Michael Polanyi) or "scientific estate" (Don Price).[6] President Eisenhower warned of the out-of-control military–industrial complex, and recently critics have paraphrased him in disparaging the predations of the medical–industrial complex (Relman, 1980).

To what extent does all this add up to technocracy or expertocracy, rule by technical experts? Some episodes have come close. Anyone searching for an extreme case would recognize as one of the most momentous technocratic affairs ever conducted the 1949 American decision to develop the hydrogen bomb. Herbert York has reflected (1976, p. 45):

> Especially considering the enormity of the issue—and most of those involved were fully aware of its enormity—the participants in the secret debate were very few: the members of the General Advisory Committee [to the Atomic Energy Commission (AEC)], the members of the AEC and a few of their staff, the members of the [congressional] Joint Committee on Atomic Energy and a few of their staff, a very few top officials in the Defense Department, and a very small group of very concerned scientists, mostly from two of the AEC's laboratories.

In the same way, and involving some of the same people who made the nuclear weapons decisions, from the beginning of the Atomic Age until recently a tight coalition of technical leaders in industry and government, almost totally deferred to by awed elected officials, made the public's nuclear power decisions. This remains the case in most countries.

A lot of science has moved from outside to inside of government. Since the 1950s and 1960s many of the advisory functions formerly filled by aca-

6. Keller, 1963; Klaw, 1968; Lapp, 1965; Polanyi, 1962b; and Price, 1965.

demic and other free-lance advisors, such as those performed by panels of the old President's Science Advisory Committee, have become integrated into the internal staff-work and standing advisory committee duties of government agencies. The State Department, Bureau of Mines, Central Intelligence Agency, and dozens of other bureaus now employ thousands of technical and technical-policy experts fulltime. Quasi-technical agencies, such as the Environmental Protection Agency and the Occupational Safety and Health Administration, have proliferated.

Similar changes have occurred in many countries and multilateral organizations. In developing nations technical people work with the political leadership to plan and direct technical education, industrialization, resource utilization, security strategy, and social research. The same in the North Atlantic Treaty Organization, Organization for Economic Cooperation and Development, and International Union for Conservation of Nature.

Establishment institutions such as the National Academy of Sciences and its Research Council have ventured—with mixed effect—into taking on ever more socially contentious assignments. In 1982 the Committee on Underground Coal Mine Safety reviewed the nineteen largest bituminous mining companies in America, ranked the companies by a variety of injury-rate indices (naming the companies explicitly), reported on labor–management, regulatory compliance, and other issues at particular mines, and drew a number of conclusions about the "persistent and large differences between the injury rates" that "could not be explained by physical, technological, or geographical conditions, but are due to factors internal to the companies."[7] At about the same time the Committee on Substance Abuse and Habitual Behavior reviewed the effectiveness of national marijuana policy and published its conclusion that "We believe that a policy of partial prohibition is clearly preferable to a policy of complete prohibition of supply and use."[8] A long social leap from the dry physical compilations of the old Committee on Elemental Line Spectra!

Quite a few technical people serve in public office, both as technical bureaucrats and as legislators and administrators. Many executives of large firms, especially those in the chemicals, metals, mining, agriculture, energy, health, and electronics industries, work their way up from R&D or engineering divisions, receive little other special training, and maintain a technical base and influence throughout their careers. M.D.'s of course play leading roles in pharmaceutical firms, medical clinics, and other health-care enterprises. Corporate technical executives direct the regulatory lobbying and other political activities of such organizations as the Pharmaceutical Manufacturers Association and the Aerospace Industries

7. NRC, Committee on Underground Coal Mine Safety, 1982, p. 5.
8. NRC, Committee on Substance Abuse and Habitual Behavior, 1982, p. 29.

Association. Other scientists work actively with such special-interest groups as the Natural Resources Defense Council, The Cousteau Society, and the United Cerebral Palsy Association.

A "science affairs community," to adopt Christopher Wright's useful label, "comprised of individuals who directly and significantly influence national science policy or the role of science in the development of national policies," exercises considerable leadership in Washington and other power centers (1964, p. 261). A number of such groups are active. Most members of these free-form influence coteries—in some areas almost cabals—are veterans of government or industry service, or are seasoned academics or medical leaders.

These aspects of milieu don't seem to me to add up to anything strong enough to warrant being called "technocracy." They are influential, certainly, and elitist in many ways; but for the most part these activities are extremely diverse, diffuse, and pluralistic. In Western countries, defense and intelligence activities come closest to being the exception. The Communist bloc countries are more technocratic in general.

We will gain insight into the concerns that give rise to suspicions of technocracy, now, as we examine some characteristics of technology and medicine.

TECHNOLOGY IN SOCIETY

Pervasiveness and Potentia *of Technology*

Ever since the emergence of *Homo faber* as "maker," technology has indissolubly been wedded to production, destruction, and prophecy. From the Middle Ages on, the builders called *ingeniators* applied their ingenuity to military tasks, as did armorers and naval architects, and by the nineteenth century they had bequeathed their technique and name to the new profession of engineering. (But, I hasten to insert, technology arises from many sources in addition to engineering.)

From the metallurgy of Damascus swords to that of armor-piercing shells, from the chemistry of gunpowder to that of cluster bombs, from early coding and signaling to the Airborne Warning and Control System, technology always has been integral to warfare. Now all aspects of national security—supply of strategic rare metals and fuels, control of weather weapons, and protection of the computerized accounts of international finance, in addition to the whole enterprise of classical military security—depend on technology. The swords-into-plowshares business of strategic arms control is thoroughly technical, involving fields from hardware reliability analysis to seismic detection to game theory.[9]

9. Technology in warfare is discussed in Zuckerman, 1967; Reid, 1969; Merton, 1973, pp. 204–209; Sapolsky, 1977; and McNeill, 1982. We urgently need a synthesizing critique of the dynamic relations between military R&D and international security.

Technology has likewise been essential in commerce, from early minerology to magnetometric prospecting, from steam engines to fuel cells, from jacquard weaving to robotics. Health care obviously has become technological, in protection against radiation and chemical hazards, in screening and vaccination, and in all of therapeutic medicine.

From the turn of the century, as engineers and scientists have become managers, the art of management has become a technical craft. Catalytic contributions were made by Frederick W. Taylor and his colleagues, working mostly at Bethlehem Iron Company, as they developed time-and-motion studies, worker fatigue analyses, differential piecework wage systems, dynamic cost accounting, and other methods into an entire school of technique and attitude. Taylor introduced many efficiencies and promoted them widely, but his excesses—he expected production workers "to do what they are told to do promptly and without asking questions or making suggestions"—became so debilitating that Taylorism itself eventually fell into disfavor. Revisions followed, though, and from that time on the daily work-lives of most manufacturing employees have been governed by quantitative production-optimizing considerations.[10] Now many aspects of administration are mathematized; linear programming and related techniques are central among the skills students pursuing Masters of Business Administration degrees are expected to acquire.

Management now heavily involves technology, and technological development heavily involves management. It is not just that communications and office work are mechanized. The cross-influence goes far beyond that, as Harvey Brooks has emphasized (1980, p. 66):

> Technology does not consist of artifacts but of the public knowledge that underlies the artifacts and the way they can be used in society. Management, insofar as it can be described by fully specifiable rules, is thus a technology, and indeed every large bureaucratic organization can be considered an embodiment of technology just as much as a piece of machinery. Thus it has been suggested that the greatest innovation in the Apollo program was not the hardware, but the managerial system. This system made possible the degree of reliability and technical discipline necessary to bring the project to a successful conclusion, through the coordination of the activities of hundreds of contractors and subcontractors. In contrast, the report of the Kemeny Commission suggests that it was the lack of similar managerial innovation in the nuclear industry that led to the mishaps of Three Mile Island and Browns Ferry.

10. Noble, 1977, is a fine critical account. Kakar, 1970, is a good biography of Taylor. Merkle, 1980, provides an international treatment. Trescott, 1983, reviews the parallel history of Lilian Moller Gilbreth's work in scientific industrial management.

Not only by physical agency but also by force of prediction in navigation, resource management, agricultural planning, and other endeavors, special technique has always conferred power—power as ability, social status, leverage, or wealth—on its masters. "*Scientia et potentia humana in idem coincidunt*," Bacon formulated it in the *Novum Organum*, "Knowledge and power meet as one."

Further, technical *potentia* enhances and socially "levels" personal ability and opportunity. Perinatal protection helps infants survive complications of gestation and birth and get a healthy start in life. Medical technologies enable early detection, correction, and compensation for sensory, dental, spinal, and other infirmities, and repair of stigmatizing disfigurements. Social technologies, such as standardized behavioral tests, complement medicine in identifying and remedying deficiencies of learning and communication. Environmental and personal technologies, such as contact lenses, insulin, and special telephones, help compensate for disabilities. Farm and transportation technologies empower rural people with physical and social mobility, as household technologies do housekeepers. Thus technologies can greatly enhance personal potential, not just political-economic might.[11]

Gloom-and-doom critics tend to ignore how technology can serve cherished humanistic ends. The ancients' polished metal mirrors and early writing systems began long lines of applied-science developments that reveal truths to us about ourselves. To construe technology as antihumanistic is to overlook benefits like this one described by John Fletcher (1983, p. 307):

> New technology allows living persons to view, study, and help the fetus in distress. The process of bonding between parents and the fetus apparently occurs much earlier in the life cycle for those parents who participate in the use of new methods. When modern parents see the fetus move, an older form of loyalty is initiated sooner. Technology does not dehumanize the fetus or the future child nearly as much as was once feared. . . . The more living persons actually *see* the fetus, by ultrasound or in any future method to be developed, the more human and valuable will the fetus become.

Photography, printing, painting, sculpting, musicmaking, video recording, and a wide range of other artistic activities involve technology, as do winemaking, bicycling, and skiing.

11. U.S. Congress, Office of Technology Assessment, 1982, discusses technology and the handicapped. Giedion, 1948, pp. 512–595, reviews mechanization of the household. Cowan, 1983, discusses "the industrialization of the home." Rothschild, 1983, pp. 79–93, waffles through the argument that "although technology has greatly eased the worst burdens of household labor, [it] has aided a capitalist-patriarchal political order to reinforce the gender division of labor and to lock women more firmly into their traditional roles in the home."

Definition of Technology

To think of technology just as fabricated objects, or tools, or machines, neglects the instrumentality of know-how and simply cannot encompass the way technology now is generated, applied, controlled, and transferred.[12]

A broadened definition was promulgated by social theorist Jacques Ellul in his 1954 classic, *La Technique*: "In our technological society, technique is the totality of methods rationally arrived at and having absolute efficiency (for a given stage of development) in every field of human activity."[13] (The qualifier, "having absolute efficiency," meant "the one best way.") The spirit of Ellul's statement has influenced most scholars' usages ever since. Ellul's "technique" usually can be taken as synonymous with "technology."

Representative of more recent formulations, most of which include reference to a science base, is Bernard Gendron's: "A technology is any systematized practical knowledge, based on experimentation and/or scientific theory, which enhances the capacity of society to produce goods and services, and which is embodied in productive skills, organization, or machinery" (1977, p. 23). Vice Admiral Hyman Rickover, the father of the U.S. Nuclear Navy, put it pragmatically: "Technology is tools, techniques, procedures, things; the artefacts fashioned by modern industrial man to increase his powers of mind and body" (1965, p. 721).

The consternations that are the subject of this book require that we define technology sufficiently robustly to convey its informational nature, span beyond mere material production, imply that technology can be possessed exclusively, and relate technology's potential to social as well as physical power. I might as well try my lexical hand.

Technology is instrumentality, employing special knowledge, that extends human effort beyond that of the unaided mind and hand. Thus technology can be a navigation system (based on elaborate grids of radio and radar transmissions), a manufacturing process (dependent on systematized, experience-based and possibly proprietary manufacturing art as well as on basic formulas and equipment), even a banking system (involving interlocking rules, codes, and procedures, some highly secret, that integrate myriad fast-changing data with physical currency and commodity flows). Not all "technique" need be included: coitus interruptus could be considered a technique, hormonal contraception a technology.

Attributes of Modern Technology

Contemporary technology has taken on many characteristics that in earlier times it didn't have, or wasn't seen as having. Its *physical scale and complexity*

12. Essays on definition of technology are McGinn, 1978, and Mitcham, 1978.
13. This definition was stated in the prefacing "Note to the Reader" of Ellul, 1964.

can of course exceed anything imaginable before, as can its *temporal scale and irreversibility*. As to complexity, one should think not just of conceptual complicatedness, but of degree of interdependence: we now live in a world in which a British electric power engineer could remark, "The coal miners brought the country to its knees in eight weeks—we could do it in eight minutes" (*Nuclear Times*, 1975, p. 3). Every clock, watch, and time-dependent instrument and process in the entire world is synchronized with a few master cesium-beam clocks, which are in turn calibrated to astronomical observations. In 1984 someone stole and posted on an "electronic bulletin board" a computer password that controlled access to the financial credit histories of ninety million Americans (Diamond, 1984). As to short temporal scale, it is not impossible for the news of a mid-air collision to be known on the opposite side of the globe before the first wreckage hits the ground. As to long temporal scale, there is a pressing need in many countries to dispose of nuclear waste whose radioactive lifetime will extend tens of times longer than any civilization has ever endured.

Technology must be seen as *disembodied knowledge and technical art*, not just as hardware. Nuclear know-how can move from MIT to Libya via the brain of a graduate student as readily as via a transported blueprint or device, and often more efficiently, given the importance of being able to adapt the techniques to indigenous systems after transferral. This raises great difficulties for control.

Major technology tends to be *embedded in sociotechnical systems*. The Green Revolution in Asia involved not only the new high-yield rice seeds but a constellation of changes in fertilization, irrigation, cultivation, pest control, and harvesting, and, beyond that, changes in taxation, credit provisions, and land tenure. Cargo containerization has had implications for everything from transport timing and pricing to warehouse security. The public now understands that energy supply is a systems problem having political, economic, and ethical complexities as important as its scientific ones.

Technology now is *developed and applied more within institutional contexts, more deliberately*. Few high-tech inventions can be expected to come from lone entrepreneurs working in their basements anymore. The Wright brothers' humble shop and Kill Devil Hills sand dunes have been supplanted by the aerospace industry–Defense Department complex's huge wind tunnels and proving grounds. R&D are planned strategically to pursue corporate or agency goals. Now we expect complex prior assessment of costs and effects before society adopts major new transportation systems, or psychosocial screening tests, or expensive medical diagnostic machines. In the United States and elsewhere these expectations have become embodied in laws, as with the environmental impact assessment requirements of the National Environmental Policy Act.

Increasingly, *technology and science intersect*. Science leads to, and supports, technological advance. Once applied, technology demands that science help solve its practical problems; it demonstrates phenomenological regularities that beg for scientific explanation; and, fiscally and otherwise, often it "carries" science within institutions. Technologies become instrumentation for scientific advance. In Eastman Kodak or NASA, then, or within the field of pharmacology, the overlap between technology and science is substantial.[14] In fields such as aerodynamics there is little distinction between the two at all. Edward Shils has emphasized that "much of the tradition which sustains contemporary technology has grown out of traditions which are not themselves integral to the technology. Computers draw on traditions from Pascal and Boole [in logic] as well as the tradition which immediately preceded Bardeen's and Shockley's work [on transistors]" (1981, p. 82). Application can quickly follow scientific discovery: in the last decade's impressive application of lasers, for example, and homogeneous chemical catalysts, DNA recombination, clinical radioimmunoassay, and fiber optics, the R&D gap and application time-lag were extremely short compared to those for similar earlier innovations.

Technology can *induce worth in things that otherwise would remain of little tangible value.* In new technological contexts, krill and other free-ranging marine biota, desolate but strategically located islands, even the moon, are being haggled over as commercial and military properties; genes are becoming subject to patenting and industrial licensing; blood, plasma, semen, and womb service (and soon, no doubt, ova and transplantable organs) are being sold as commodities; and weather-harvesting techniques are putting price tags on the very clouds. As ever, technology begets wealth and power, and wealth and power tend to beget technology. The classic factors of production—land, labor, and capital—have been broadened extraordinarily: to arable and mineable land have been added maricultur-able seas and mineable seabeds, the polar ice-caps, navigable airspace, geostationary satellite orbits, and allocable frequencies of the radiospectrum; to strong backs and skilled hands have been added robots, computer systems, and pedigreed bacteria; and to durable capital have been added arcane manufacturing and military know-how, elements of time and timing, information, information about information, and the "negative capital" of burdens that accompany production, such as facilities made brittle and radioactive by long exposure to nuclear material.

Thus technology, which must be viewed as comprising disembodied knowledge and technical art as well as hardware, deeply intersects "pure science"; it is embedded in sociotechnical systems; it now can carry unprecedented physical and temporal scale, complexity, and irreversibility; it tends to be developed and applied

14. Cyril Stanley Smith, 1967, describes the technology/science interaction in metallurgy. Geballe, 1981, describes it in solid-state electronics. Illinois Institute of Technology Research Institute, 1968, and Batelle Columbus Laboratories, 1973, review the interaction for ten modern innovations. Nathan Rosenberg, 1982, surveys the interaction in several areas 1860–1914.

within institutional contexts, more deliberately; and it can induce worth in things that otherwise would remain of little tangible value.

So deep do these attributes of technology run that it will require the elements of every chapter of the rest of this book to trace the ethical consequences and suggest means for exerting social choice and control.

Technological Determinism?

"The things we call 'technologies' are ways of building order in our world," Langdon Winner observed. He pointed out the way New York master builder Robert Moses designed overpasses on Long Island expressways to have low overhead clearance, so as to exclude buses and their generally lower-class and minority riders, especially those on their way to his favorite project, Jones Beach. Winner cited other examples, such as Louis Napoleon's directing Baron Haussmann to cut cannon-shot-straight boulevards through Paris as a precaution against street riots like those of 1848. He correctly concluded (1980, p. 127):

> Consciously or not, deliberately or inadvertently, societies choose structures for technologies that influence how people are going to work, communicate, travel, consume, and so forth over a very long time. In the processes by which structuring decisions are made, different people are differently situated and possess unequal degrees of power as well as unequal levels of awareness. . . . Because choices tend to become strongly fixed in material equipment, economic investment, and social habit, the original flexibility vanishes for all practical purposes once the initial commitments are made.

In his extended critique, *Autonomous Technology*, Winner sensitively traced the origins of the notion that "somehow technology has gotten out of control and follows its own course, independent of human direction" (1977). Like Winner, I believe that pessimistic theme was illegitimately reinforced by the French writer Jacques Ellul. Perhaps more strongly than Winner, I think Ellul perpetrated a disservice with the often-quoted but outrageous assertion in his 1954 book, *The Technological Society*, that "everything which is technique is necessarily used as soon as it is available, without distinction of good or evil."[15] While *the spreading of new ideas* is almost inevitable, surely *the application of ideas* is not.

What is true is that technological systems, once developed and deployed, exert a "soft determinism," as economist Robert Heilbroner has called it (1967). Countless future electrical engineering options are decided once a country makes its choice between an alternating (AC) and direct current (DC) electricity system (Thomas Hughes, 1983). A nuclear electric econ-

15. Ellul, 1964, p. 99. Christians and Van Hook, 1981, carries essays on Ellul.

omy, once entered into, unavoidably requires security safeguards such as personnel background investigation, nuclear materials accounting, and undercover surveillance. Building the Paris *Métro*, the West German *Autobahn*, and the U.S. Interstate Highway System channeled many future options of land use and regional economics, and influenced innumerable secondary and derivative options.

It makes no sense, though, to attribute to technology a mind of its own. "In the management of our affairs we have too often been bad workmen," Peter Medawar has admonished, "and like all bad workmen we blame our tools."[16] The slogan of the 1933 Century of Progress Exhibition—"Science finds; industry applies; man conforms"—was a premature and unnecessary capitulation.

As I will describe throughout this book, many means are available for analyzing, evaluating, tailoring, and controlling technologies. There still is every reason to believe we can shape technologies to desired ends. And *contra* many writings on these problems, I would insist that practical concern is never about some looming abstraction, TECHNOLOGY, but about specific applications of particular technologies.

MEDICALIZATION OF LIFE

Many core aspects of life have become "medicalized." Doctors now routinely apply their craft to everything from abortion and sex-change to bedwetting, midlife crankiness, and dignity in dying. "One significant form that the process of medicalization has taken is the increase in the numbers and kinds of attitudes and behaviors that have come to be defined as illnesses and treatment of which is regarded as belonging within the jurisdiction of medicine and its practitioners," sociologist Renée Fox has pointed out.[17]

> Although sin, crime, and sickness are not related in a simple, invariant way, there has been a general tendency in the society to move from sin to crime to sickness in categorizing a number of aberrant or deviant states to the degree that the concept of the "medicalization of deviance" has taken root in social science writings. The sin-to-crime-to-sickness evolution has been most apparent with respect to the conditions that are now considered to be mental illness, or associated with serious psychological and/or social disturbances. These include, for example, states of hallucination and delusion that once would have been interpreted as signs of possession by the Devil, certain forms of

16. Medawar, 1972, p. 135. The theme of "blaming technology" is given popular treatment in Florman, 1981.

17. Fox, [1977] 1979, pp. 468–483; this is an excellent essay.

physical violence, such as the type of child abuse that results in what is termed the "battered child syndrome," the set of behaviors in children which are alternatively called hyperactivity, hyperkinesis, or minimal brain dysfunction, and so-called "addictive disorders," such as alcoholism, drug addiction, compulsive overeating, and compulsive gambling.

More than ever, physicians are being asked to play Aesculapian Notary roles. Doctors certify the moment of birth and the moment of death, define gender and aspects of sexuality, approve syphilis-free marriageworthiness, authorize sick leave and disability payment, counsel on advisability of having children, testify as to fertility, paternity, and sexual violation, and judge many questions of social normalcy and sanity. It is not at all evident that medical schools prepare young physicians for these roles.[18] Scientists often are partners in this work, as when, in assessing hormonal, genetic, and other disorders that are not yet routinely analyzable, physicians consult biochemists and other researchers for sophisticated chromosome, hormone, and enzyme studies. Diagnosing and treating hyperactivity in schoolchildren or anorexia nervosa in teenagers often requires that doctors collaborate closely with psychologists.

If one believes, as I certainly do, that labeling of persons carries social implications, one may be dismayed that it was only in the mid-1970s, and after years of acrimonious internal debate, that the American Psychiatric Association exonerated homosexuality from being classified as a "psychopathic personality" trait in its official *Diagnostic and Statistical Manual*.[19] The way technical labeling can create social categories is exemplified in the extreme by the case of Soviet psychiatric diagnosis of schizophrenia, which is considered abusively overdiagnostic (too inclusive) by most Western psychiatrists: "The system created a category, first on paper and then, with training, in the minds of Soviet psychiatrists, which was eventually assumed to represent a real class of patients and which was inevitably filled by real persons."[20]

In most countries physicians work with other specialists to arrive at determinative definitions of "legally blind" and "[il-]legally intoxicated," and they help judge in individual cases whether the labels apply. These roles of typing people and certifying their attributes to employers, teachers, insurers, courts, and the government are coming under increased public scrutiny, as they should.

"Psychiatrists exert power in so many different ways that it is difficult to keep them all in mind," the Emory professor of law and psychiatry Jonas

18. On the medical "gatekeeper" role, see Deborah Stone, 1979.

19. For a recounting of the controversy, see Bayer, 1981.

20. Walter Reich, 1981, p. 71; this volume, edited by Bloch and Chodoff, provides a good overview of psychiatric ethical issues.

Robitscher observed in his treatise, *The Powers of Psychiatry* (1980). Robitscher listed fifty-one ways psychiatrists exercise authority, including: decisions concerning involuntary administration of medication, electroshock treatment, and behavior modification for hospitalized patients; determination of competency to stand trial; diagnosing subjects as sexual psychopaths or defective delinquents under appropriate statutes and, by doing so, diverting them from the criminal-justice to the mental health system; testimony to determine which of two divorcing parents should have custody of a child; evaluation for appropriateness of vasectomy or plastic surgery; evaluation for job separation on the basis of mental instability or psychiatric problems; and evaluation for honorable as compared to dishonorable military discharge. In recent years the Institute of Medicine has advised the State Department on psychological strategy for coping with terrorists in foreign countries, and it has been consulted by the Secret Service on ways of profiling potential political assassins (Institute of Medicine, 1981).

Psychiatrists hold dual—if I may, sometimes schizoid—responsibilities: helping their clients within the haven of privacy, confidentiality, and consent, while at the same time protecting society from violence or disruption by those clients. Psychiatrists have not been eager to recognize or resolve the conflicts of policing versus therapy, or to address the larger context of their work. "Under the comfortable illusion of being a value-free discipline in a period that still accepted the concept of a value-free science, psychiatry ignored the immense ethical and moral implications of its activity," the ethicist and psychiatrist Willard Gaylin has said. "We were therapists, ministers of medicine, not of the soul; servants of the body and its extension, 'the mind'—divorced from both ethical and political concerns."[21]

The issue of defining death starkly exemplifies the overall relation between medicine and the state. In 1981, to replace guidelines based on breathing and heartbeat that have been made obsolete by the development of technological life-support systems, after long deliberation the President's Commission for the Study of Ethical Problems in Medicine and Biomedical and Behavioral Research recommended that all states adopt a Uniform Definition of Death Act. The proposed Act reads:[22]

> An individual who has sustained either (1) irreversible cessation of circulatory and respiratory functions, or (2) irreversible cessation of all functions of the entire brain, including the brain stem, is dead. A

21. American Psychiatric Association and The Hastings Center, April, 1978, p. 2. The duality-of-roles problem is also discussed in several chapters of Bloch and Chodoff, 1981, and Stone, 1984.

22. U.S. President's Commission for the Study of Ethical Problems in Medicine and Biomedical and Behavioral Research, July 1981.

determination of death must be made in accordance with accepted medical standards.

Both the American Medical Association and the American Bar Association dropped their own, similar proposals and endorsed that of the President's panel. This definition has now been adopted in many states. The Act still defers to "accepted medical standards," as it must, but it has influenced those standards.

Thus a great range of issues that once were the province of family, community, and church have been ceded to the authority of physicians. The challenge, which this book will address in a variety of ways, is to ensure that doctors' Aesculapian Notary authority, and indeed all medical actions, reflect society's values.

Three

The Complexion of Science and the Social Sciences

All attitudes, mores, institutional structures, and controls relating to the social activities of *science-doing* and *science-using* are affected by the complexion of *science-knowing*. The task of this chapter is to encapsulate the features of scientific knowing, then those of social-scientific knowing. As with the sketch of technology in the preceding chapter, my intentions are to raise issues, state my own approach, and set the foundation for the rest of the book.

CONSTRUCTION OF SCIENTIFIC KNOWLEDGE

Many of the public's frustrations with science, and many public disputes, stem from failure to appreciate the basic nature of scientific fact. The underlying questions are fundamental ones. How does a tentative conjecture become a Fact? Why do scientists disagree with each other so much? Why can't objective use of the scientific method spare us from uncertainty? How can it be that truth changes over time? When apparent facts come into conflict, how is the conflict resolved?

In the early part of this century science took on too positivistic a cast. Science was portrayed as being a campaign to discover "objective" facts about nature, using The Scientific Method to ferret out "natural" laws. Mainstream scientists have supplanted this with a much more dynamic conception, which I will now outline. Maturation on these points has been stimulated by advances in the history, philosophy, and sociology of knowledge, challenged by the increasingly public promotion and evaluation of technical work, and enriched by senior scientists' reflections.[1]

Scientific Knowing

Facts are not hidden answers "out there" waiting to be found, but concepts that are invented, shaped, and fitted together into conjectural models and maps. "The objects of scientific inquiry," Jerome Ravetz has aptly put it, are "classes of intellectually constructed things and events, the evidence for whose properties is derived from particular experiences. . . . A scientific problem is thus incapable of having a solution which is 'true'. Rather, the solution will be assessed for adequacy."[2] Elegant and powerful though they may be, scientific laws are simply constructs of the mind. Grand generalizations like Newton's laws of motion, Ohm's law of electrodynamics, or the laws of thermodynamics come along only rarely. "Objectivity" is, as I will argue shortly, a troublesome notion.

Even observations are relative. The weight of an apple is measured relative to tare-weights calibrated to standards, and the weight depends on how high relative to the center of the earth's gravity the weighing is done, and on whether the apple-on-scale assemblage is accelerating relative to the center of the earth—in short, on the entire freshman physics kit. Circularity is unavoidable: theory is based on and always must be consonant with observation, but it is impossible to observe without making at least implicit reference to some theory or its embodiment in a scheme of description or measuring device. It means nothing, scientifically, to speak of that apple's weight without making reference to Newtonian or other theories of "weight-ness."[3] A directive to "go out and determine whether Agent Orange causes birth defects" leaves open a lot of questions about exactly what "Agent Orange" chemically was as experienced (dioxin impurities?), what the character of exposure might have been (respiratory, dermal? prolonged, brief?), what constitutes "causing" any of the anomalies (which?) called "birth defects," and by what evidential clues and logic all these questions should be sorted out.

Science is eternally provisional, and is a matter of consensus: the scientifically true is what scientists endorse as being true. No more, no less. "In science, 'fact' can only mean 'confirmed to such a degree that it would be perverse to withhold provisional assent,'" Stephen Gould put it (1981, p. 35).

---

1. Some of the changes in philosophy and history of science 1950–1977 are described by Toulmin, 1977.

2. Ravetz, 1971, p. 153; this is a classic exposition.

3. The observation/theory distinction and many other such matters are instructively reviewed in Achinstein, 1968, and Suppe, 1977.

For a candidate fact to be admitted to the canon of certified knowledge it must pass a process of authentication. Michael Polanyi's description is classic (1962b, p. 68):

> There are no mere facts in science. A scientific fact is one that has been accepted as such by scientific opinion, both on the grounds of the evidence in favour of it, and because it appears sufficiently plausible in view of the current scientific conception of the nature of things. . . . Science *is what it is*, in virtue of the way in which scientific authority constantly eliminates, or else recognises at various levels of merit, contributions offered to science.

Even deep theory remains contingent and is valued mainly for its directive power. In 1861, two years after he finally had published *The Origin of Species*, Darwin mused: "I believe in Natural Selection, not because I can prove in any single case that it has changed one species into another, but because it groups and explains well (as it seems to me) a host of facts in classification, embryology, morphology, rudimentary organs, geological succession and distribution" (1861). J. J. Thomson, the discoverer of the electron, in 1907 said similarly: "From the point of view of the physicist, a theory of matter is a policy rather than a creed; its object is to connect or coordinate apparently diverse phenomena, and above all to suggest, stimulate and direct experiment" (1907, p. 1).

Any item of newborn knowledge must be justified against existing knowledge, either by fitting consistently with it or by refuting it. This justificatory obligation on the inventor of an idea is one of the central burdens and strengths of science. For a candidate piece of science, intellectual filtering and quality control begin in the laboratory of its origin, are applied again when the putative new knowledge is subjected to scientific editorial review before publication, are applied again when the originators or others integrate the new information into more definitive statements, and are applied whenever the facts or theories are reviewed in critical articles, conferences, monographs, policy-relevant assessments, grant or patent decisions, or technicolegal disputes.

Each field fosters its own criteria of probity, but these usually remain informal and only weakly codified. For example, the instructions for submission of papers to the *Journal of the American Chemical Society* specify that elemental analyses are acceptable if they fall within 0.4% of the theoretical values calculated for the compound; promulgated, on the basis of long experience that this is almost always "good enough," by the editorial board of the world's most prestigious chemical journal, this sets a norm for laboratory practice. For the most part criteria "just grow" as craft knowledge, become woven casually into educational instruction, and only rarely are specified in journals and other authoritative sources. Nonetheless, criteria of factuality are firmly applied. In this regard science is relentlessly egalitarian. Being a distinguished scientist in no way exempts one from having to meet standards and argue fairly.

Scientific knowledge is evaluated by two kinds of criteria: analytic and predictive power, and fruitfulness in leading to new conjectures and realizations. Ultimately, good science is science that "works": science that can *make sense* of phenomena and *predict* with consistency and accuracy and generality what will happen in the physical, biological, and social world. It must also try to *explain*, but we tend to evaluate explanations against their power to lead to prediction. Some sciences, built on analysis of patterns and trends, predict better than they explain.[4]

One of the most remarkable things about science is how strongly its internal quality-control norms are maintained, even though they are inculcated only informally and for the most part informally enforced. Complete, honest experimental records and specimens are expected to be kept; findings must be submitted for publication in a form that will allow exact independent duplication of the observations; misrepresentation of any kind is abhorred; tests of statistical significance must be met. But rarely is there a rulebook.

Science is accumulative and intellectually progressive. Its physical accretive character is evident in its extraordinary reference collections of fossils, beetles, seeds, shells, alloys, tree-ring cores, genetic data, astronomical photographs, geologic maps. The reach of its data registries is enormous. *Index Medicus* continuously digests 3,000 medical publications from around the world. *Chemical Abstracts* covers some 12,000 different journals and lists references to 350,000 previously unregistered chemical compounds every year (Abelson, 1982). This physical cumulation is a mere indication of the intellectual cumulation that occurs.

All new findings must be "placed" in respect to various concept-history streams, with which they either meld or remain immiscible. Thus, as Stephen Toulmin put it, "Every concept is an intellectual micro-institution" (1972, vol. 1, p. 166). It is not only answers that accumulate—questions and problems do, too.

In respect of this concept-history placement, scientific argument is by nature historiographic. Research reports define the traditions they take off from ("Continuing the search for _____," or "Contrary to _____'s finding that _____, we found that _____"). They trace the geneology of themes ("This experiment was meant to test, in our own case of _____, the notion of _____, recently developed from _____'s classical theory by _____"). They recall constellations of assumptions and methodological limitations ("The method we used was a variation of the _____ assay to detect _____, known to be an indicator of _____"). They suggest anal-

---

4. For discussion of "explanation," see Nagel, 1961. Regarding the related notion "retrodiction" as applied to the physical sciences, see Kitts, 1977, and, as applied to historiography, Veyne, [1971] 1984.

ogies ("As _____ is known to do _____, our similar system _____ seems to be doing _____, which, by the way, suggests that in _____ other cases _____ might turn out to do _____"). And they portray the evolution of ideas ("Earlier notions of _____, supplanted by the concept of _____, now can be recast as _____, thus accounting for the apparent inconsistency _____, which explains why _____, and this ties _____ to _____, all of which leads to our new conjecture _____, which [if we get our next grant] we hope to explore by doing _____").

Science is progressive in its very essence. A college freshman can easily come to understand more about some aspects of biology or physics than his professors' professors. Today's modestly educated technician can improve on the laboratory practice of yesteryear's Nobelists.[5]

I can't detour through the controversies surrounding Karl Popper's *Logic of Scientific Discovery*, Thomas Kuhn's *Structure of Scientific Revolutions*, and other such works here; that would require a book in itself. The debates have been stimulating, although they have remained amazingly uncontaminated by reference to what transpires in real laboratories. These arguments, which have been more unsettling to philosophers than to scientists, have addressed such issues as what scientific rationality consists in, whether it is useful to distinguish "normal science" from "revolutionary science," whether the act of "proving" involves anything other than falsifying untrue propositions, how logic of proof is related to psychology or sociology of discovery, and how science "progresses."[6]

Overall, science pursues two precariously consonant objectives at once: preserving orthodox knowledge and carefully building upon it, all the while striving to supersede it with more powerful knowledge. This need not be seen as paradoxical. Science is, as I have emphasized, eternally provisional. New knowledge need not necessarily negate the knowledge it supplants, but may translate it conceptually or encompass it and extend it through a larger domain.

Thus, to summarize these fundamentals: *Rather than being hidden answers "out there" waiting to be found, facts are concepts invented, shaped, and fitted together into conjectural models and maps. Science is eternally provisional, and is a matter of consensus: the scientifically true is what scientists endorse as being true. Scientific knowledge is judged by two overall criteria: analytic and predictive power, and fruitfulness in leading to new conjectures and realizations. Science is accumulative and intellectually progressive, preserving orthodox knowledge and*

5. Laudan, 1977, despite some shortcomings, is a stimulating commentary on problem-posing and the meaning of "progress" in science. Along with it should be read Hull, 1979; Lugg, 1979; Butts, 1979; Jarvie, 1979; and Doppelt, 1983.

6. General sources include Kuhn, 1970; Lakatos and Musgrave, 1970; Doppelt, 1978; Latour and Woolgar, 1979; Mulkay, 1979; Gutting, 1980; and Cartright, 1983. Readable expositions on change in geology, which is a superb case-example, are McKenzie, 1977; Glen, 1982; and Hallam, 1983.

carefully building upon it, all the while striving to supersede it with more powerful knowledge.

To ensure high quality, scientists employ elaborate collective procedures of screening, cross-checking, and criticism, which I will describe in later chapters. Of course, review procedure per se can't guarantee correctness. In the long run, science that "works"—solves conceptual or practical problems—gets adopted and its developers rewarded; science that doesn't work gets denounced or, much more commonly, is simply ignored.

Norms of Science

It is not easy to list the norms of science. I have described several aspects of scientific stance above. Talcott Parsons said the "basic norms" were "empirical validity, logical clarity or precision of the particular proposition, logical consistency of the mutual implications of propositions, and generality of the 'principles' involved" (1951, p. 335).

A much-quoted depiction is that by sociologist of science Robert Merton, who described the "ethos of modern science" as being guided by four imperatives: *universalism* ("truth-claims, whatever their source, are to be subjected to preestablished impersonal criteria: consonant with observation and with previously established knowledge"); *"communism"* (scientific knowledge becomes public property, and the process of its certification is communal); *disinterestedness* (truth-seeking is to be valued above personal aggrandizement, and constantly must be subjected to peer accountability); and *organized skepticism* ("temporary suspension of judgment" and "detached scrutiny of beliefs" are methodologically and institutionally mandatory) ([1942] 1973, pp. 268–278). Later Merton cited the "potentially incompatible values" of *"originality*, which leads scientists to want their priority to be recognized, and . . . *humility*, which leads them to insist on how little they have been able to accomplish" ([1957] 1973, p. 305). To these sociologist Bernard Barber added *rationality* and *emotional neutrality* (1952, pp. 84–88).

The Mertonian norms are suggestive but are open to criticism.[7] They are broad, and Merton has not defined them very precisely, either originally or subsequently. The phrase "organized skepticism" evokes the ethos of science perfectly, as does the term "universalism," but neither is directive of how skepticism should be organized or universalism ensured. "Communality" does apply in general, but a lot of science having military or industrial potential is held very closely by its discoverers until exploitative advantage has passed. Given all the ways scientists scramble for recognition, support, and reward, "disinterestedness" is elusive. Probably the best way to describe it is to say that scientists urge themselves and their col-

7. Two critiques, neither satisfactory, are Mulkay, 1977, pp. 105–106, and Barnes and Dolby, 1970.

leagues to perform research detachedly and dispassionately, to confront each idea or experiment as though for the moment no particular outcome matters, in a kind of willing suspension of intuition—"scientific agnostic doubt," Michael Polanyi called it (1962a, p. 274). But a devoted researcher can hardly help championing one theory over another, one exciting near-solution over less promising alternatives. And in science accomplishment does very much matter, as in all human endeavors: intellectual humility before the Universe, yes, but not excessively before one's grant review committee.

Philosophy, "The" Scientific Method, and "Objectivity"

Scientists are, in a strong sense, applied philosophers. However, as John Ziman has observed, "The average scientist will say that he knows from experience and common sense what he is doing, and so long as he is not striking too deeply into the foundations of knowledge he is content to leave the highly technical discussion of the nature of Science to those self-appointed authorities the Philosophers of Science. A rough and ready conventional wisdom will see him through" (1968, p. 6).

Philosophizing is too often viewed by scientists as being either abstruse or effete. In 1965 medical statistician Austin Bradford Hill began his presidential address to the newly formed Section of Occupational Medicine of the Royal Society of Medicine by disclaiming, "I have no wish, nor the skill, to embark on a philosophical discussion of the meaning of 'causation.'" But then through his entire, elegant speech he went on to discuss the concepts of strength of causal association, consistency and specificity of evidence, temporal order of events, biological dose-response gradients, experimental variation, analogical reasoning, and plausibility and coherence of explanation, that are the everyday philosophical tools of his trade (1965). Scientists should be urged to discuss such fundamentals more vigorously within their professions, with their students, and with public leaders. Many public controversies, such as those over how to assess chemical risks, hinge on just such philosophy.

There are two kinds of assertions that I think ought to be avoided, because they aggravate the public's confusion over science used in analyzing public problems: claims of exclusive access to the scientific method, and claims of objectivity.

To the extent that the phrase tempts anyone to assume that mastery of method ensures mastery of scientific thinking, "the scientific method" misleads. It is hard to say exactly what scientific method comprises, other than taking fastidious care to isolate variables, observe carefully, and restrain one's declarative sentences. As physicist Percy Bridgman explained, "I like to say there is no scientific method as such, but rather only the free and utmost use of intelligence" (1950, p. 278). Medical researcher Peter Medawar, who views science as "a logically articulated structure of justifiable beliefs about nature," also expressed it flexibly: "The scientific method is a potentiation of common sense exercised with a specially firm determination not to persist in error if any exertion of hand or mind can deliver us from it" (1969, p. 59). Each field, of course, has its methods of experimental craft.[8] More than by any grand all-purpose Method, scientists are guided by styles of creative approach, by favored laboratory techniques and test subjects, and by their communities' largely tacit standards governing admissibility of evidence, repeatability and precision of observations, and statistical and logical persuasiveness of analyses (Nash, 1963, pp. 157–179 and 320–326).

Any imperious claims to objectivity in fact-finding should be suspect. This antiquated notion once was used to connote emotional detachment of observer from object observed, to mean that measurements could be independently verified by other observers, to imply that facts were allowed to fall wherever they might. There is a residue of meaning to the thought: people wishing to gauge an apple's weight may, instead of hefting it in hand, agree to objectify the determination by reading the apple's weight on a scale; but questions still remain about which scale, used in what manner, corrected for what inaccuracies, and so on, and these judgmental problems are compounded if the weighing of the apple is replicated at other places and times, or if the issue is "What is the weight of an average apple?"

A claim to being as objective as possible is like a claim to being as virginal as possible—the idea communicates only obliquely, and the term can easily be abused. To their credit, most special-interest groups have stopped bragging that they develop "objective" opinion on environmental, health, or defense issues.

The fatal challenge to the notion of objectivity is: By what criteria could one possibly confirm this self-claim to correctness, other than by subjectively judging the persons, reputations, and procedures involved? Judging evidence on its internal logical merits is a separate issue, and does not hinge on anything that could be called objectivity. Performing analyses by committee is no guarantee of detachment, either. The term "objectivity" should be abandoned.

Best to aim for studies that can be defended as being precisely defined, strictly and honestly reported, state-of-the-art, reliable, independently confirmable, judiciously balanced, broadly critiqued, authoritative. Although these are based on abstract tenets of evidence and logic, they remain socially evaluated criteria for which there can be no nonsocial substitute. The best index of these qualities in an analyst or organization is a sustained record of involving the most competent people, using the most reliable procedures, performing studies as dispassionately as possible—and

---

8. For discussion of scientific investigation as craft, see Polanyi, 1962a, and Ravetz, 1971, pp. 75–108.

actually having turned out to have been powerfully perceptive and predictive. Nothing beats a record of being correct.

COMPLEXION OF THE SOCIAL SCIENCES

"Thou shalt not sit/ With statisticians nor commit/ A social science," W. H. Auden admonished in his poem "Under Which Lyre"—and how Auden's unease would have been compounded had he foreseen the way the social sciences now consort with statistics!

In an inflammatory 1975 *New York Times Magazine* article (September 28, 1975) entitled "Knowledge Dethroned," Robert Nisbet, a Columbia University sociologist, complained:

> I have run into no responsible social scientist recently who will not, at least in the confidence of his study, concede that during the nineteen-fifties and nineteen-sixties we were giving, and demanding the right to give, advice as giants when we were still pygmies. We didn't have then and we still don't have more than a very small amount of knowledge when it comes to problems of recession and inflation, urban social problems, the behavior of electorates, and scores of other exceedingly complex matters on which so many of us were not merely willing but eager to give profusely of our counsel.

Nisbet's criticism moved the *American Scholar* to solicit a symposium of responses entitled "Social Science: The Public Disenchantment" (1976). Asking, "Is Robert Nisbet wrong in saying that scholars have much to answer for?" Harvard government professor James Q. Wilson conceded that "Although there were some conspicuous exceptions . . . many scholars who should have known better acquiesced in the process of allowing eager politicians to convert bad ideas into dubious policy—either by participating, very much on the periphery of this process, or by failing to say, loudly and frequently enough, that the emperor had no clothes" (p. 359). A review of anthropology and public policy in 1980 concluded that "Anthropologists have not had significant, visible impact on policy formulation in any major domestic or international policy area" (Hinshaw, 1980, p. 516). In the federal domestic budget-cutting spree of 1981, while the National Science Foundation as a whole was granted an increase, the budget for the agency's social, behavioral, and economic sciences programs was slashed because the White House deemed this research to be "of relatively lesser importance to the economy."

What's the problem? Why Auden's crack, Nisbet's wail, Wilson's misgivings? Are the theories and facts generated by the behavioral and social sciences—which include major aspects of anthropology, geography, demography, linguistics, economics, history, psychology and psychiatry, political science, and sociology—inadequate to their self-assigned tasks? Are these disciplines inherently doomed to remain inadequate, or are they just immature? Or, are they accomplishing all that can be expected? How do they interact with other technical endeavors?[9]

We must now discuss the social sciences, because more and more they are coming into intersection with other technical fields, and because they provide insights into people's values and feedback on the social impacts of technology.

Over recent years social-scientific research has moved toward the other sciences both in topics covered and in methods employed, and this has yielded some fruitful collaborations. Mutual recognition has increased. In the late 1960s the social sciences gained full status in the National Science Foundation and the National Academy of Sciences, and a Nobel Memorial Prize for Economic Sciences has been awarded annually since 1969.

But, as economist Herbert Simon has reminded us, "Newcomers the social scientists are not" (1980, p. 72).

> The first mortality tables were published by John Graunt in 1662, and the first calculations of life annuities based on such tables were carried out by the astronomer Halley in 1693, just 7 years after he assisted Newton with the publication of the *Principia*. Cournot's pioneering work in mathematical economics, containing a deep analysis of the phenomena of imperfect competition, appeared in 1838, a generation before Maxwell wrote out the basic equations governing electricity and magnetism.

Profundity in the social sciences has not, however, been limited to quantitative investigations, as is evident from the contributions of Adam Smith, Karl Marx, Sigmund Freud, Max Weber, Claude Lévi-Strauss, Jean Piaget, and Noam Chomsky (although several of these have engaged in parametric thinking).

Is Social Research Different?

Briefly, for heuristic purposes, I would like to probe for differences between social research and physical research, keeping in mind economist Amartya Sen's protest that to define social science as meaning "something very like natural science with the same techniques applied to social matters

9. Nowadays there almost seem to be more analyses of the social sciences' woes than there are analyses of society by the social sciences. Recent critiques, not all of them readable, from a variety of points of view are Eisenstadt, 1976; Brown, 1977; Howe, 1977; Dallmayr and McCarthy, 1977; Papineau, 1978; Rabinow and Sullivan, 1979; Geuss, 1981; Cronbach, 1982; Gergen, 1982; NRC, Committee on Basic Research in the Behavioral and Social Sciences, 1982; Callahan and Jennings, 1983; Lloyd, 1983; Outhwaite, 1983; Runciman, 1983; Sabia, 1983; Thurow, 1983; Haan et al., 1983; and Kaplan, 1984.

. . . is rather like defining aerial warfare as techniques of naval warfare applied in the air!" (1983, p. 96).

Surely the social sciences should be considered different from the physical and biological sciences. Although the objects of study in all fields resist ultimate understanding, when the objects under investigation are the quirks, whims, vices, words, and actions of human beings and groups of human beings, some of them long dead, analysis inevitably must be—to understate—multifactorial, imprecise, and structurally complex. Replication of observations and experiments will be difficult. Experiments analogous to the core experiments of, say, inorganic chemistry may well be forbidden ethically. The very act of observing may influence the phenomena under observation.

Causal relationships may appear multitudinous, as in this hyperbolic example from sociologist John Seeley (1963, p. 60):

> The cause of delinquency, "other things being equal," is any one or more of the following: poor street lighting, alleys, immigration, paternal infidelity, differential association, neurotic acting out, broken homes, the American income distribution, lack of alternate meaningful activities, advertising and display, failure to nail down prized objects, the slum, the ecological organization of the American city, materialism, its opposite; preoccupation with one's worth as a person, the law itself, the absurdity of society or the human condition; the want of religion, the nuclear family, the political system which seeds crime which needs as a training ground prisons and reformatories; schools that engage few or no loyalties, the perversity of the individual delinquent, or his parents, or theirs; psychological ignorance, the unconscious wishes of those who deplore the activity or condemn the actors.

But equally multitudinous, for that matter, are the causes of weather, heart disease, and other classic subjects of science.

Mechanisms of causality may be hard even to postulate. "Cutting and trying" may be more fruitful than theorizing. As Harvard economist Marc Roberts pointed out in a superb essay, "Social science is much closer to engineering or meteorology than it is to mathematical physics or astronomy" (1974, p. 57).

Often it is complained that the social sciences lack theoretical foundations. Many of the grand conceptions on which eighteenth- and nineteenth-century social theorists built their fame have been consigned to history, and some, such as Marx's theories about the consequences of capitalism, have remained untested and probably are, in most particulars, untestable. But so has Darwin's version of Darwinism been consigned to history as being neither fully correct nor testable in its particulars. Besides, some sciences as "hard" as hydrodynamics are powerfully useful despite being based more on models and approximations than on unified theory (Birkhoff, 1960, elegantly establishes this point). Just as an apple's weight reflects a notion of "weightness," social-science observation reflects theory, even if the theory is only weakly developed. As Milton Friedman insisted in his *Essays in Positive Economics*, "A theory is the way we perceive 'facts,' and we cannot perceive 'facts' without a theory." Sounding just like Darwin and Thomson as quoted earlier, Friedman went on to say that "Economics as a positive science is a body of tentatively accepted generalizations about economic phenomena that can be used to predict the consequences of changes in circumstances" (1953, pp. 34 and 39).

A challenge perennially raised to the social sciences is whether they can either explain or predict anything the way, say, physics and geology do. The social sciences do find regularities among variables that help make sense of observed phenomena. Often social scientists hold, as Marc Roberts put it, that "to 'explain' something is simply to tell a story about what has happened that makes it plausible that events should have occurred as they did" (1974, p. 49). To compare with the physical sciences, this is not much different from N. R. Hanson's remark on the discovery of the neutrino that "Physical theories provide patterns within which data appear intelligible" (1958, p. 90). Social scientists devise explanatory notions the same way physical scientists invent concepts such as angular momentum.[10] Reviewing Émile Durkheim's studies of suicide, Toby Huff made it clear that "Durkheim's innovation was a *theoretical* innovation achieved through a series of *conceptual* innovations which reduced the causes of variations in rates of suicide from near-infinity to a finite and parsimonious set" (1975, p. 254).

> In effect, Durkheim was postulating "imaginary" conceptual categories which if found to exist—i.e. corroborated—would explain the variations in rates of suicide "as a matter of course." The categories of "egoism," "altruism" and "anomie" are of course theoretical ideas, creations of reasoning, not "sense-data." They were not "observed" and "measured" in the first instance in any of the usual sensationalist or empiricist connotations of those operations.

But can the social sciences predict? This is less easily affirmed. Economists, wags say, have predicted fifteen out of the last nine recessions. Demographer H. H. Winsborough of the University of Wisconsin recently testified:[11]

> A demographer at the University of Pennsylvania warned of the end of the baby boom in the mid-1950's, another at The University of Chi-

---

10. Laudan, 1983, carries provocative essays on explanation in psychiatry.

11. U.S. House of Representatives, Subcommittee on Census and Population, 1982, p. 440.

cago signaled the end of rising birth rates in developing countries, a colleague in my own group and his co-author in Agriculture described the turn-around in growth between metropolis and open country, and a demographer at Cal Tech described the changes in mortality at older ages.

However, Calvin Beale, the renowned Agriculture Department expert Winsborough was referring to, has observed that American demographers were "surprised" by at least six major trends of the 1970s: they overestimated the birthrate, in part because they did not sufficiently anticipate deferral of marriages and childbearing, and the increase in abortions; they underestimated life expectancy, not anticipating reduction in cardiac mortality; they underestimated the decline in household size; they failed to predict the population migration to the West and South; they failed to forecast nonmetropolitan growth, the rate of which tripled; and they failed to anticipate the large increase in illegal and refugee immigration (Population Reference Bureau, 1982, p. 3).

From circumstances and premonitory events psychologists can predict trends in people's behavior, just as volcanologists can predict trends in volcanic activity. Economists and political scientists can predict some consequences of decisions and events. Sometimes this is mere trend charting; other times it involves theoretical modeling. Discontinuities such as urban riots or crimes of passion may be no more precisely predictable than, say, earthquakes are. The social-science establishment did not usefully predict the 1973 OPEC oil boycott, the dethronement of Shah Pahlavi Reza, or the recent relapse of the Brazilian economy. Both the analytic disciplines and our use of them in decisionmaking seem to be faulty. As in all the sciences, social and behavioral predictions need to be specified precisely and modestly. Predictive failures need to be critiqued.

Echoing Comte's *"Savoir pour prévoir, prévoir pour pouvoir"*—"Knowledge for prediction, prediction for enablement"—social scientists usually emphasize, as Gunnar Myrdal did, that "The social sciences have all received their impetus much more from the urge to improve society than from simple curiosity about its working" (1958, p. 9). This sets a Scylla/Charybdis problem. If researchers devote themselves to solving social ills, and if they work intimately with their subjects, they may find it hard to manage personal biases and avoid influencing, and being influenced by, the people they are observing. Their facts may well become fused with values. If, on the other hand, researchers keep rigorously distant and theorize broadly, they end up talking only to themselves and, in Marc Roberts's phrase, publishing mere "recreational mathematics."

Social-scientific findings may turn out to be counterintuitive. This may discomfit patrons. As Kenneth Prewitt, president of the Social Sciences Research Council, noted, "Very often the research results do not accord with the way those in responsible positions believe the world to be, or wish it to be" (1981, p. 5).

> Demographers show that education or housing or the health plans of local governments are thwarted by unanticipated decreases, or increases, in crucial age groups; the economic concept of externalities obliges us to take account of the environmental costs of an imperfectly regulated market system; sociologists and economists are mapping a large "hidden economy and informal labor sector," the presence of which argues against any simple solutions to declining industrial productivity; political scientists anticipate the deleterious effect on political parties of various reforms in party rules; evaluation research has shown the faulty assumptions of programs as diverse as housing subsidies, bonus systems, health-cost containment schemes, and curriculum reform.

But the natural sciences also generate intuitively unsettling findings. Ecological research has revealed that the soil of the Amazon Basin, far from being robust as the jungle's fecund appearance leads one to assume, is thin and delicate, and once stripped of its protective canopy easily deteriorates. Medicine has found that many babies, such as those lacking enzymes to digest lactose, are harmed by their own mother's milk.

Fact-finding is value-laden in the social sciences as in the other sciences, but, again, probably in a more immediate and obvious way. Listen to a few veteran commentators. Gunnar Myrdal (1969, p. 55):[12]

> A "disinterested" social science has never existed and, for logical reasons, can never exist.... The only way in which we can strive for "objectivity" in theoretical analysis is to expose the valuations to full light, make them conscious, specific, and explicit, and permit them to determine the theoretical research.

Sociologist Edward Shils, who has referred to social scientists as "the contemporary equivalents of the *philosophes*" ([1977] 1980, p. 459):

> Even the deepest and most abstract of contemporary social scientists are, in no pejorative sense, publicists. They are writing most frequently about contemporary society, and, even when they declare their allegiance to the ideal of an ethically neutral science of society, they quite consistently do not allow that to stand in the way of their ethical judgment and political partisanship with respect to the society which they study.

And similarly, philosopher Ernest Nagel (1961, p. 489):

---

12. For similar comments by economists see Robinson, 1963, p. 14, and Heilbroner, 1973; by a political scientist, Taylor, 1967.

Although the recommendation that social scientists make fully explicit their value commitments is undoubtedly salutary, and can produce excellent fruit, it verges on being a counsel of perfection. . . . The difficulties generated for scientific inquiry by unconscious bias and tacit value orientations are rarely overcome by devout resolutions to eliminate bias. They are usually overcome, often only gradually, through the self-corrective mechanisms of science as a social enterprise.

I cannot imagine how a researcher could claim pure dispassionateness when living with kibbutzim in an anthropological study, surveying the attitudes of migrant workers' children, or theorizing over childbeating, homophobia, or nuclear bargaining strategy. Social scientists face severe challenges on this score when they serve as advisors and advocates.

Probably more quickly than in the physical sciences, tentative social-science findings tend to get picked up by the press and cited in social debate. Long before the larger social-science community and other experts have had a chance to critique a report it can be used as ammunition for arguments beginning, "A recent study of [cocaine use, television violence, whatever] has shown. . . ."

Social-scientific conjectures are quite vulnerable to being abused politically. Joseph Ben David has stated how this can put the researcher in a difficult position (1979, p. 32):

The social scientist faces several alternatives: he may stand up and attempt to disavow the misuse of his ideas, sometimes at considerable cost to his scientific career and at risk of failure (because he is not likely to be able to compete with professional propagandists in enlightening the public); he may simply ignore the whole thing, keep out of political controversy and, by default, become an accomplice in the misuse of his work; or he may submit to temptation and reap the benefits of popularity and social success by sanctioning, explicitly or implicitly, the intellectually dishonest use of his ideas for some practical cause.

Thus distinctions between the social sciences and the physical and biological sciences mostly are differences in degree rather than in kind. But I think the differences are large.

The social sciences embrace some of the most heterogeneous, varying, self-conscious, mischievous, "nonideal" subjects in the universe—Ourselves—and in this they face daunting problems in observing without influencing; in replicating observations; in working intimately enough with their subjects to gain detailed insight, while preserving analytical detachment; in proving that effects stem from particular causes; in constructing mechanistic explanations; in managing fact–value entanglements; and in defending their findings against abuse.

Problems of Quality

It is telling that when the Reagan administration threatened to chop the budgets of the social-science divisions of the National Science Foundation, the affected researchers felt obliged to defend their very existence, not just their productivity under this one funding source. Testimonials by the Social Science Research Council, the National Science Board, and other bodies sounded disappointingly like those made thirty-five years previous during the debate over whether the newly established NSF should fund these fields at all.[13] They didn't sell themselves well. For success stories they pointed to methodologies such as demographic projection, voter and consumer polling, standardized educational testing, economic indicators, and geopolitical analysis; to conceptualizations such as game theory and theory of organizations; and to applied research on industrial innovation and on child development. Some of these in my opinion are dubious successes, and some, such as cost–benefit analysis and game theory, until recently have been nurtured more by engineering and operations research than by mainline social science.

In congressional testimony NSF director John Slaughter emphasized that "the research supported by NSF is not directed toward proving or disproving particular positions regarding social policies or actions." But his disclaimer was in effect countered by other witnesses' testimonials, as when University of Wisconsin economist Ralph Andreano tutored the committee:[14]

> Many government programs and taxes have disincentive effects with respect to work effort, saving, and investment. This statement is the basis of what has come to be called "supply side economics." Basic theory and methods for measurement of these effects have been developed by economists with NSF funding.

There is no question that the social sciences potentially are "relevant." But conspicuously absent from the NSF defense was documented attestation by grateful non-social-scientist beneficiaries that their problems had actually been solved by particular NSF-sponsored social research. The NSF chemistry division would have no difficulty getting industrial chemists and other users to testify, with impressive specificity, to the problem-solving relevance of its programs. There *are* grateful users, many of them in the business world, and it seems to me that non-social-scientist testifiers could easily have demonstrated to the Congress and administration how many

13. U.S. Congress, Congressional Research Service, 1979, pp. 113–144; U.S. House of Representatives, Subcommittee on Science, Research and Technology, 1981a; Social Science Research Council, 1981; and Branscomb, 1981. For an update, see Holden, 1984.

14. U.S. House of Representatives, Subcommittee on Science, Research and Technology, 1981a, p. 667.

fruitful ideas about industrial innovation, productivity, marketing, and so on derive from and are nourished by social-science research.

In 1982 a National Research Council committee published an apologia, *Behavioral and Social Science Research,* that concluded:[15]

> Basic research in the behavioral and social sciences has yielded an impressive array of accomplishments, and there is every reason to expect the yield from future research to be as great. . . . The benefits of basic research are seldom if ever predictable in advance. . . . The coupling between basic research in the behavioral and social sciences and its applications to public policy is significant and growing, but it is also inherently loose, uncertain, incomplete, and often slow.

Notice the committee's prime example: "Perhaps the single most important information-gathering invention of the social sciences is the sample survey, developed over the course of the past 50 years by statisticians, sociologists, economists, political scientists, psychometricians, and survey specialists" (p. 64). Along with some misadvised former political campaigners, I remain unimpressed. The record is hardly one a field of basic research should be proud of. A 1977 "Survey of the American Professoriate," perpetrated by two high-ranking sociologists, was embarrassingly poorly designed (Lang, 1981). About that time the American Statistical Association's "Pilot Assessment of Survey Practices and Data Quality in Surveys of Human Populations" found (Bailar and Lanphier, 1978, p. 13):

> Fifteen of 26 Federal surveys did not meet their objectives, four because of poor design, four more because of failure to implement plans for probability sampling, and the remaining seven because of a combination of serious technical flaws. These technical flaws included high nonresponse rates, failure to compute variances or computation of variances in the wrong way, inclusion of inferences in the final report that could not have been substantiated by the survey results, no verification of interviewing, and no data cleaning.
>
> Seven of the ten non-Federal surveys did not meet their objectives, one because of poor design, four more because of failure to implement plans for probability sampling, and the remaining two because of a combination of serious technical flaws. ·

A 1981 National Research Council panel, tiptoeing through the problem, gave survey research a mixed review and recommended many procedural improvements.[16] Clearly the field has a long way to go. And this is "the

15. NRC, Committee on Basic Research in the Behavioral and Social Sciences, 1982; better than the summary Part I is the supplementary Part II, which is an exciting collection of specialist essays.

16. NRC, Panel on Survey Measurement of Subjective Phenomena, 1981.

single most important information-gathering invention of the social sciences"?[17]

Outlook

Generally as to the use of social research, the 1982 National Research Council review approvingly quoted Harvard education policy expert Carol Weiss (Weiss, 1977, p. 534):

> Evidence suggests that government officials use research less to arrive at solutions than to orient themselves to problems . . . to help them think about issues and define the problematics of a situation, to gain new ideas and new perspectives . . . to help formulate problems and to set the agenda for future policy actions. And much of this use is not deliberate, direct, and targeted, but a result of long-term percolation of social science concepts, theories, and findings into the climate of informed opinion.

Which recalls the sardonic closing paragraph of John Maynard Keynes's *General Theory of Employment, Interest and Money* (1942): "Practical men, who believe themselves to be quite exempt from any intellectual influences, are usually the slaves of some defunct economist." (But then, we may wonder, where do economists get their ethical systems? The far-from-defunct economist George Stigler answered recently, "Wherever they can find one." Besides, he said, "Economists seldom address ethical questions as they impinge on economic theory or behavior" [1981, pp. 167 and 145]. This is a problem for later in the book.)

My view is that the social sciences are, by and large, still in their adolescence. They have built descriptive bases, established elaborate data-collection and -analysis systems, generated some provocative theories and useful methods, and helped solve some social problems. Their evaluations of the social impact of technological developments provide reflexiveness that is essential to judicious innovation. Now they are supplanting too-simplistic early methods (culture-bound IQ tests, equilibrium economic analyses) with more sophisticated approaches.

Probably the social sciences' deepest accomplishment has been their inculcation in popular understanding, through education and scholarly activity, of the way notions such as social structure, subculture, symbolic communication, roles, investment, mobility, and feedback make it envisionable "how" individuals and groups of people act and interact, and thus help make systematic sense of social forces, possibilities, and changes.

17. An essay that conveys a sense that large-scale survey methods can be rigorous is Tanur, 1982. A morally skeptical essay is Sievers, 1983.

Cross-fertilization between social research and other technical endeavors promises rich developments. Collaboration between psychologists, endocrinologists, and neurophysiologists on such problems as child development has been very productive. Perceptual psychologists and biophysicists symbiotically are revolutionizing our neuroscientific understanding. Social-science researchers are helping reform health care, as medical sociologists analyze doctor–patient interactions, decision analysts reveal to doctors how they in fact go about making clinical decisions, biostatisticians and operations researchers evaluate the efficacy and risks of medical diagnoses and treatments, economists analyze how fee structures affect medical practice, psychologists develop techniques for modifying lifestyles, and as human ecologists from various disciplines analyze how tropical people's living habits affect their relationships with parasites. Human-factors engineers, using a blend of psychology, design, and common sense, are helping improve nuclear powerplant control by investigating operator procedures, fatigue, stress, expectancy or "set," skill adequacy, communications, and controlroom design.[18] Social scientists are revealing the dynamics of technical innovation, and, through social and environmental impact assessments, are helping us anticipate the consequences of change. Thus we stand to benefit greatly as the social sciences become integrated into the overall scientific effort and contribute to the reflexivity and social tailoring of technical endeavor.

18. U.S. Nuclear Regulatory Commission, 1980.

Four

Professionalism and Responsibility

Much of society's ethical change in recent years, especially that concerning technical work, has entailed the "institutionalization of trust."[1] Expectations of social responsibility have been shifted toward collectives of experts, rather than remaining with experts as individuals, and attempts have been made to codify those expectations. This has given rise to the ethos of professionalism and professional responsibility that we now will examine.

What is professionalism? What is involved in "professional judgment"? In what ways can and should technical people, by applying their professional craft, contribute to society's wellbeing and decisions of conscience? To what extent should they be expected to take initiative and bear special guardian responsibilities? Why? How adequate a motivation and guide is "responsibility"?

PROFESSIONALISM

The classic professions of law, clergy, and medicine originated at European centers in the late Middle Ages as high-class guilds. Architecture, engineering, dentistry, and accounting evolved behind them, and nursing and pharmacy followed. By the mid-1800s such scientists as chemists and geologists began to seek recognition as professionals.

Emerging fields sought professional status to connote higher training and seriousness, to distinguish professional geologists from unreliable ore

---

1. Perrucci and Gerstl, 1969, among others, develops this notion.

assayers and rockhounds, professional archaeologists from curio collectors, professional ornithologists from birdwatchers. (Now the term, as happens to most descriptors of quality and exclusiveness, has been appropriated by "professional" boxers, cooks, and banjo players, simply to connote recognized, remunerable skill.)

Technical professionals can be characterized as groups of experts who hold special educational and skill qualifications, who are granted near monopoly in providing services or rendering technical judgments, and who exercise largely autonomous control over their membership and practice.

The formality and strictness in society's control of professional practice tends to parallel the directness with which the work affects members of the public. Dentists are subject to much more direct and detailed societal constraint than particle physicists are.

Although most research scientists now are considered professionals, their professionalism usually is not definable or enforceable in the same way that that of obstetricians or certified public accountants is. Scientists have acquired the status in part through self-promotion ("a professional is anyone others are willing to treat as a professional"), in part through their association with the classic professions, and in part, perhaps, through the professorship status of their academic members.

Medical Professionalism

The profession of medicine coalesced in important new ways as a result of Abraham Flexner's 1910 critique, *Medical Education in the United States and Canada.*[2] Since then medicine has struggled to become more scientific; developed institutional quality controls; excluded a variety of charlatanisms; worked out relations with nursing, physical therapy, pharmacy, optometry, and other related fields; and skeptically explored such "fringe" practices as hypnotism, chiropractic, and acupuncture.[3]

Nowadays to practice medicine in the United States doctors must meet elaborate societal and peer controls. They must have graduated from a medical school accredited by the Liaison Committee on Medical Education, which is sponsored jointly by the American Association of Medical Colleges and the American Medical Association. They must have completed an internship and, usually, residency accredited by the Coordinating Council on Medical Education, a council that represents the interests of the American Association of Medical Colleges, the American Board of Medical Specialties, the American Hospital Association, the American Medical Association, and the Council of Medical Specialty Societies. They

2. Flexner, 1910. For history surrounding the Flexner report, see Kaufman, 1976, pp. 154–182.

3. A classic though dated overview is Freidson, 1970. General histories of American medicine are Duffy, 1976; Bordley and Harvey, 1979; and Starr, 1982.

must have earned a license from a state medical board, based on testimonials as to sound moral character and on state examinations that usually are derived from exams prepared by the National Board of Medical Examiners, itself a broadly representative medical body. To qualify for specialty practice, such as dermatology or pediatrics, they must pass additional examinations, and in some specialties must recertify periodically. Doctors usually belong to national and local medical societies. Two-thirds of American physicians belong to the American Medical Association, a voluntary membership society and lobbying organization. Physicians' work is subject to accountability by their employing institutions, such as clinics, military institutions, and health maintenance organizations; by third-party payers, such as insurance companies; and by client lawsuits. Over 375,000 doctors practice in the United States.

Medicine is science-based. Medical school applicants must take courses in basic biological and physical sciences and must demonstrate competence in them on the Medical College Admission Tests. Early in medical school students have to sweat through omnibus tours of subjects like biochemistry. The basic themes of science of the human body are what are important. Not many doctors a few years out of residency can solve acid–base buffer equations or recall the Krebs cycle the way they could in medical school; but they cannot possibly serve as clinicians without knowing how to interpret changes in body fluid pH and liver chemistry signs. Few physicians trust themselves to perform detailed statistical calculations; but they could hardly read medical journals if they didn't understand basic statistical reasoning. Some physicians, of course, conduct medical-scientific research themselves.

Beyond therapeutic medicine, professional health care—an enormous and diffuse endeavor that can be taken to include not only medical diagnosis and cure but also environmental and personal hygiene, nutrition, sexual adjustment, adaptation to aging, and many other matters—remains deeply fragmented.

Engineering Professionalism

Engineering developed with less unity than medicine did, although by the turn of the century the great engineering schools had developed core curricula and engineers from different specialties were collaborating on major projects.[4]

Engineering was deeply influenced by the Progressive movement early in the century. As an outgrowth of its positivistic triumphs, engineering took on a special complexion (Layton, 1971, p. 57):

---

4. Excellent histories of aspects of engineering are Layton, 1971, and Sinclair, 1980. A general critical history of modern engineering has yet to be written.

By 1920, the philosophy of professionalism had become something of an obsession with engineers. Three themes served at once to express and to encourage this new ideology. In speeches delivered in the period 1895 to 1920 before major engineering societies, presidents and others in the vanguard of professionalism portrayed the engineer in glowing terms. They saw him as the agent of all technological change, and hence as a vital force for human progress and enlightenment. Secondly, such men drew an image of the engineer as a logical thinker free of bias and thus suited for the role of social leader and arbiter between classes. Finally, these speeches indicated that the engineer had a special social responsibility to protect progress and to insure that technological change led to human benefit.

Much of that tone still pervades the field, although now, as I observe them, most engineers prefer to avoid overt, direct sociopolitical activity.

Engineers are practical problem solvers. They specialize in such fields as chemical, ceramic, textile, petroleum, mechanical, nuclear, electronic, metallurgical, astronautic, industrial, or agricultural engineering. Skills diverge early in school: a student of mining engineering at the Colorado School of Mines has little in common with a biomedical engineering student at Duke, and neither would be able to perform the cost engineering for a new oil refinery. The majority work in business and industry, and because they are involved deeply in corporate planning, construction, and operation, in many firms they are close to management. Indeed, engineers often become managers. Many top corporate and public-utility executives have worked their way up from shirtsleeves engineering beginnings.

To practice engineering a person simply has to hold a diploma from a school accredited by the Accreditation Board for Engineering and Technology, an independent board governed by representatives of eighteen major engineering societies. Engineers may earn a state license by passing a series of state-administered examinations. Fewer than half of American engineers currently are licensed. However, the substance of public engineering projects, such as port facilities or dams, almost always is required under local, state, or federal law to be approved by a licensed engineer.

Engineering, like medicine, is based on scientific principles. But because many aspects of engineering deal with the behavior of materials or systems quite distant from the "ideal" situations of basic science, what is important is systematic, scientific *approach*: what works, works; and practice is as important as theory. A lot of engineering is high craft. Engineers are good at analyzing how things work and modifying them to work better, whether tankers, television systems, or artificial human knees. For example, engineers have applied mathematical operations-research techniques, which were developed during World War II for optimizing merchant marine ship loading and routing, to optimizing the collection, storage, and routing of blood in hospital and regional blood-bank systems.[5]

Because the specifics of practice are so diverse and have originated from so many disparate sources, engineering has grown up fragmented. The profession still has difficulty holding its many specialties together. A few strong specialty societies and national umbrella organizations are active, but in the United States no unified national engineering society has ever emerged. On the broad scale, Robert Perrucci and Joel Gerstl's characterization of engineering as being "a profession without community" was not too much an exaggeration (1969).

In their practice engineers generally must adhere to formal and informal standards established by specialist organizations, and they must respect federal, state, and local design and performance codes (such as electrical, sanitary, and building-construction codes). Professional organizations, such as the Institute of Electrical and Electronics Engineers, provide substantial guidance. Within firms engineers are subject to cross-check by other engineers. Lawsuits are always a possibility. In the United States as many as 1,300,000 people are identified by training and work as engineers.

Applied-Science Professionalism

Scientists are situated differently from both physicians and engineers. Formal education and licensing apply mainly as conditions for employment and support, and much less as conditions on the kind of work scientists are allowed to perform. Many quality criteria are applied to scientists' work, of course, as employers and colleagues review it. Researchers must obey a wide range of civil regulations regarding handling of chemicals, testing of explosives, disposal of wastes, monitoring of radiation, protection of human subjects, and importation of botanical materials. And they and their institutions are subject to lawsuit for negligence and other transgressions.

In applied fields in which scientific reliability has to be assured to employers and public authorities, certification may be instituted. The American Society for Microbiology reviews candidates' education and experience and administers examinations to certify microbiologists for work in clinical laboratories, dairies, waterworks, and breweries. The American Board of Industrial Hygiene similarly certifies members' competence.

Applied social scientists seem only recently to have begun to confront the sense of their own professionalism. Some, such as clinical psychologists, in most states are required to earn state licenses to practice. For the most part economists, anthropologists, sociologists, and other social sci-

5. Miser, 1980, provides a succinct summary of this example and of the field.

entists simply trade on their reputations, and any free-lancer having a college education may try to join in.

Codes of Professional Ethics

The archetypal code of specialist ethics is, of course, the Hippocratic oath. Recently an alternative to that nonspecific, idealistic oath has been provided by the American Medical Association's ten-point Principles of Medical Ethics, an improved, more detailed, but still idealistic code (1979). I must say I find that a dismaying number of the AMA principles have only to do with financial matters; that some, such as stipulation 3.50 that "a physician should not use unscientific methods for treatment nor should he voluntarily associate professionally with anyone who does," can be abused to prevent change; and that some, such as Section 5.62's insistence that "the patient has no legal right" to possess his own medical record, exemplify what is wrong with the American practice of medicine. Nonetheless, the code provides much more concrete guidance than the Hippocratic oath did.

Many technical societies have promulgated codes of ethics. The codes may address relations between practitioners and clients, practitioners and employers, practitioners and research subjects, practitioners and other practitioners, practitioners and the general public. They may address issues of research protocol, publication, compensation and fees, referrals, advertising, confidentiality, whistleblowing, and malpractice. Few of their guidelines can be considered sanctionable.

I must agree with a recent staff project of the American Association for the Advancement of Science, which, after reviewing many technical organizations' ethics programs in detail, concluded (Chalk, Frankel, and Chafer, 1980):

> As a group, the ethical statements adopted by the societies demonstrated a marked preference for dealing with ethical issues on a general and abstract level. . . . Very few of the societies' statements provided a clear basis for establishing priorities between two or more rules which, although not inherently inconsistent, in practice may present the scientist or engineer with conflicting obligations. . . . The legal status of the societies' enforcement of their ethical rules, including the application of sanctions and support actions, is not clearly established. . . . Formal complaint procedures, safeguards respecting the rights of all parties, and sanction and support actions rarely are available and even more rarely used. The societies appear to share a common assumption that complaints involving ethical concerns or code violations should be handled in an informal and private manner. As a result, formal decisions in response to individual complaints are

rare and are not publicized to the members of the profession or to the general public.

The process of drafting codes of ethics is not without worth. It can sensitize members to issues. And if developed and endorsed by membership, codes can serve as reference ideals when controversy arises. Codes are more useful if they are complemented by complaint review procedures, ombudsman services, and education and certification programs.[6]

PROFESSIONAL JUDGMENT

In this age of specialization there is little alternative to depending on the kind of judgment rendered when a radiologist reads an x-ray, a statistician critiques a census study, or a petroleum geologist recommends where to drill exploratory wells. Many technical decisions, such as routine medical diagnoses, require considerable skill but are so standardized as to need only routine quality control. At the other extreme some decisions, such as those of strategic defense planning, transcend experience, hinge heavily on extratechnical factors, and must be based on wisdom and authority far beyond anything covered by standard professionalism.

Of special difficulty, and of special interest for this book, are nonroutine judgments about uncertain facts within uncertain value contexts. In such situations high intellectual craftwork may be involved, implicit social valuations may be promoted knowingly or unknowingly, and intuitions based on experience may be pivotal.

Carcinogenicity Testing as an Illustration of Professional Judgment

To take an applied-science example illustrating the "craft" element, consider chemical carcinogenicity assessment in small mammals. All over the world every day such tests are performed, mostly on mice and rats. Yet how tedious and tenuous they are. I once worked for several unforgettable months in a laboratory where cancerous mice were being milked to study whether mammary tumors were transmitted mother-to-daughter via viruses in the milk, which impressed me indelibly (1976, p. 53):

> The day-to-day practical difficulties of managing animal experiments almost have to be endured personally to be appreciated: feeding uniformly and on schedule, keeping the cages, tanks, or stalls clean, controlling the ambient temperature, protecting the animals from disease and vermin, keeping records, collecting specimens, arranging matings, performing deliveries, shielding the young from harm, conducting autopsies, and in general taking every possible measure to ensure

6. Unger, 1982, is a good discussion of codes and ancillary matters in engineering.

that whatever unusual effects are observed will not be due to uncontrolled or unnoticed variation in experimental conditions but will be solely and compellingly attributable to the hazard under suspicion. And this perhaps with many hundreds or thousands of smelly, squealing, squirming, defecating, biting animals.

Discretionary judgments have to be made about which genetic strains of rodents to use (sensitive inbred varieties, or robust hybrids?); how many animals to involve; what chemical doses to subject them to, and through what modes of exposure (respiratory? dietary? cutaneous?); how to feed, mate, and protect the general health of the animals; when and how to take fluid and tissue specimens and make other measurements; whether to follow through several generations; when and how to kill the subjects (in infancy, so as to identify birth defects? in robust middle age? at natural death?); how to autopsy and prepare tissues for postmortem analysis; what pathologies to examine the specimens for; how to define precancerous and cancerous conditions (should nonmalignant tumors be counted?); and so on and on. Every one of these strategic and tactical decisions carries costs and possible informational payoffs. Every one is judgmental.

Biases accumulate as toxicologists favor one test condition over another, as they discard results they feel are spurious, and as they decide when to stop testing. Biases also accumulate as pathologists excise bits of brain, liver, ovary, and other tissues; dissect, fix, stain, and section the tissues into thousands of gossamer slices; select samples of the sections for examination; microscopically scan sample areas of the selected sections; and draw conclusions. And, since the more one looks for pathological signs, the more one finds, where to stop?

Interpretive judgments then have to be made in critiquing overall experimental design, aggregating the neoplastic finds into groups, adjusting statistics for background cancers and for noncancerous illnesses and deaths in the test and control populations, and drawing implications for humans. The only logics available for extrapolating from the higher doses used in the animal tests to the typically very much lower exposures of human concern, and from rodents to humans, are trend-projection and analogy, neither of which is very powerful. Too, just because an animal test system is successful at detecting some confirmed human carcinogens does not mean that it will necessarily respond to all human carcinogens. Intuitions about carcinogenic mechanisms and the human body's resiliency lead different scientists to adopt different "models" of cancer induction and consequently to arrive at differing estimates of human hazard. Beyond the science, overt societal value judgments eventually are made, of course, when experts and authorities weigh all the risk and benefit factors and decide what to do about the cancer risk.

Rodents aren't people. Nonetheless, these tests provide essential clues. Risk analogies can be drawn among structurally similar chemicals. Some animals seem to be good predictors of certain types of human carcinogens. Corroboration can be secured from epidemiological studies. Although animal testing embodies a scientific approach, the judgments involve craft experience as much as textbook science, and must continue to do so at least until we are able to explain mechanistically *how* carcinogens cause cancer.

In the past few years constructive criticism of animal testing has led to major improvements. Laboratory animal husbandry has become much more sophisticated and standardized. Industrial laboratories have taken more initiative in refining test procedures. Organizations such as the Society of Toxicology have become more active. Federal agencies responsible for toxic chemicals have been revising their approaches. And the Office of Technology Assessment has prepared a sound report on *Technologies for Determining Cancer Risks from the Environment.*[7]

What I would emphasize by this example is that often the only science available for application to important societal problems may be quite uncertain and rather crudely empirical. If this is so for animal carcinogenicity testing—the assays by which countless foods, pharmaceuticals, industrial chemicals, and pollutants are screened for their potential to cause the dominant dread disease, cancer—even shakier are the foundations of other assessments, such as those for reproductive and behavioral toxicity. Much hinges on expert judgment accrued through specific experience, which in turn hinges on assumptions, handling of uncertainties, embedded value judgments, and other underlying, or flanking, issues. Much of this judgment cumulatively becomes embodied in norms of professional practice. Chapter Seven will return to these issues, which I call "meta-analytic" considerations.

Judgment in Other Technical Fields

All applied fields cultivate their own norms. "Sound engineering judgment" is the by-phrase of that profession. The term connotes, among other things, conservative or cautious design, such as embodying "safety factors" that would prevent structural failure even under strains beyond the worst expected. A vast body of engineering know-how has become standardized. The American Society for Testing and Materials (ASTM) coordinates voluntary development of consensus standards for an enor-

---

7. U.S. Congress, Office of Technology Assessment, June 1981. Task Force of Past Presidents of the Society of Toxicology, 1982, is a succinct critique of methodology. NRC, Committee on the Institutional Means for Assessment of Risks to Public Health, 1983, reviews inferential judgment in toxicology and epidemiology. Whittemore, 1983, essays on fact/value issues in environmental toxicology. Except for its warping of the term "science policy," McGarity, 1979, usefully reviews this issue in regulatory context. U.S. Office of Science and Technology Policy, 1984, is a recent federal review.

mous variety of matters, from railroad axle dimensions, to noise testing procedures, to diesel fuel composition, which it details in a 48-volume compendium now covering some 6,600 standards. Committees of the Institute of Electrical and Electronics Engineers (IEEE) similarly develop standards for materials and procedures. These ASTM and IEEE standards, and others like them on which engineering practice depends, are so firmly established that in 1964 a federal court ruled: "Because of the heavy reliance of federal, state, and municipal governments upon ASTM for specifications, the Society may be regarded as an essential arm or branch of government, and its acts may be entitled to immunity from the antitrust laws accorded governmental acts."[8]

Such institutionalized engineering judgment is vulnerable in three aspects. Like consistency for consistency's sake, it may simply standardize the inadequate—although certainly the programs I have reviewed seem solid. It may be deemed collusion: in 1979 federal courts assessed $9,900,000 in antitrust damages against the American Society of Mechanical Engineers for letting a large manufacturer's interests dominate boiler-control standards.[9] And it may leave well-intentioned standard setters liable to lawsuit for having abetted lax protection against hazard.

In social-science studies, judgment pervades every step from the definition of problems, to the design of studies, to the conveyance of findings to managers and public leaders. Studies of the phenomena called "intelligence," behavioral "stress," "minimal brain dysfunction" (hyperactivity in children), "urban" problems, "poverty," and "unemployment" are value-laden, very sensitive to definition, and coupled to larger conceptions of humanity and society. In 1982 Harvard professors William Alonso and Paul Starr still found it important to point out the judgmental dimensions of such seemingly cut-and-dried matters as the U.S. Census:[10]

> [Census] statistics are commonly presented and accepted as neutral observations, like a weatherman's report on the day's temperature and atmospheric pressure. This view is too simple because official statistics do not merely hold a mirror to reality: they are the product of social, political, and economic interests which are often in conflict with each other; they are shaped by presuppositions and theories about the nature of society; and they are sensitive to methodological decisions made by complex bureaucracies with limited resources.

8. Memorandum opinion and order sur application of American Society of Testing and Materials, regarding whether ASTM should be considered co-conspirator in *U.S.* v. *Johns-Manville et al.*, miscellaneous no. M-2699, slip opinion (E.D. Pa., July 20, 1964).

9. *Hydrolevel Corp.* v. *The American Society of Mechanical Engineers, Inc.*, 635 F.2d 118, 1982. The episode and related issues are described in Rueth, 1975; Green, 1979; and Sinclair, 1980, pp. 214–219.

10. Alonso and Starr, 1982, p. 29. For a symposium on these issues, see Alonso and Starr, 1985.

Even routine research activities, such as psychologists' adopting statistical significance at the 95% confidence level as adequate for the test of a hypothesis, reflects judgment about how important it is to the researchers, and to their patrons or clients, to be correct.

Medical judgment grows within specialties, subject to occasional broader scrutiny by the profession and outsiders. Biases begin with the very conceptions of "health," "disease," "illness," and "cure." Ultimate values are embodied in the Aesculapian Notary judgments described earlier: certifying birth and death, sanity and insanity, wellness and sickness, fitness and unfitness. Judgment is, of course, the essence of the healing art.[11]

Doctors usually defer to prevailing medical practice. Prevalence, however, can't guarantee "best." Recent years have brought systematic evaluation of efficacies, benefits, risks, and costs of clinical procedures, in the *micro* for individual clients and in the *macro* for society as a whole. (Chapter Seven will discuss these analyses in detail.) An example of constructive review is the Consensus Development Program sponsored since 1977 by the National Institutes of Health, which brings together medical research scientists, practicing physicians, consumers, and others to review the safety and efficacy of medical and dental drugs, devices, and procedures. Task forces develop consensus statements and recommendations. Several dozen conferences have been held on such topics as burn care, mass screening for colo-rectal cancer, the Pap test for cervical cancer, hip joint replacement, and cesarean delivery. The Conference reports, which are not regulatory but advisory, are published in remarkably clear language and carry critical comments by organizations and individuals.[12]

In many areas doctors have developed consensus guidelines. The American Academy of Pediatrics Committee on Fetus and Newborn conducted a major assessment of whether administration of oxygen to premature infants increases the risk of inducing retrolental fibroplasia (a retinal pathology that can lead to blindness), and how this risk is counterbalanced by the oxygen's general benefit in preventing cerebral damage (Supplement to *Pediatrics* 57, no. 4, 1976). A Joint National Committee involving thirty large biomedical organizations published a consensus report on *Detection, Evaluation, and Treatment of High Blood Pressure*.[13] Recently an ad hoc group of experienced clinicians drafted a statement on physicians' responsibility toward hopelessly ill adult patients (Wanzer et al., 1984). Both to standardize its members' practice and to inform the public, the American College of Obstetricians and Gynecologists has issued *Guidelines on Pregnancy and Work* and *Guidelines for Perinatal Care*.[14]

11. Feinstein, 1967, is a classic overview of medical judgment.

12. The convening agency is the Office for Medical Applications of Research, National Institutes of Health.

13. *Archives of Internal Medicine* 144, pp. 1045–1057, 1984.

14. Available from the American College of Obstetricians and Gynecologists, 600 Maryland Avenue SW, Washington, DC 20024.

PROFESSIONAL RESPONSIBILITY

Enigmatically in his notebooks of the early 1500s Leonardo da Vinci indicated that he had an idea "how by a certain machine many people may stay some time under water. And how and why I do not describe my method of remaining under water . . . and I do not publish or divulge these, by reason of the evil nature of men who would use them to practice assassinations in the depths of the sea by stoving in ships and sinking them together with the men in them."[15] But though he may have kept the idea of the submarine secret, Leonardo unrestrainedly promoted himself to Duke Sforza of Milan as a one-man defense R&D consultant for devising advanced mortars and siege machines, pumping water out of moats, tunneling under fortifications, and contriving "various and endless means of offense and defense."

Ambivalence of intention and action has typified most such heroic/tragic cases ever since. World War II raised these issues to concern as never before. The Bomb problem was paramount of course, but, then as now, was a special case. How far can the notion of responsibility extend?

One postwar incident illustrates how little things have changed since da Vinci. Norbert Wiener, the MIT mathematician who during the war had brilliantly developed ways of directing antiaircraft gunfire and generalized it into a broad theory of communication and control, was asked in 1946 by the Boeing Aircraft Company for some information relating to missile guidance. In a letter reprinted in the January 1947 *Atlantic Monthly* he replied, "The interchange of ideas which is one of the great traditions of science must of course receive certain limitations when the scientist becomes an arbiter of life and death." Judging that "I cannot conceive a situation in which such weapons can produce any effect other than extending the kamikaze way of fighting to whole nations," Wiener said, "I can only protest *pro forma* in refusing to give you any information concerning my past work." Dramatically he vowed: "I do not expect to publish any future work of mine which may do damage in the hands of irresponsible militarists" (1947). But then Wiener went right on in 1948 to publish his classic text *Cybernetics*, whose concepts, along with those of his related publications, made possible a wide range of technologies from military missiles to manufacturing automatons. From his speeches and writings about rockets and "automatic factories" it is clear that Wiener fully anticipated these uses of his work.[16]

After 1945 the proposition of special responsibility was raised repeatedly. But the response of most scientists was bland. In the March 1948 *Bulletin of the Atomic Scientists* Robert Oppenheimer argued that "The true responsibility of a scientist, as we all know, is to the integrity and vigor of

15. My translation of the notebooks reprinted in Richter, 1970, p. 274.
16. Wiener, 1948, 1956; and Heims, 1980.

his science," and concluded that the suggestion that a scientist should "assume responsibility for the fruit of his work . . . appears little more than an exhortation to the man of learning to be properly uncomfortable" (1948, p. 67). In the same issue of the *Bulletin* Percy Bridgman, the Harvard physicist who had just been awarded the Nobel Prize, wrote, "Let the scientists, for their part, take a long-range point of view and not accept the careless imposition of responsibility, an acceptance which to my mind smacks too much of appeasement and lack of self-respect" (1948, p. 71). Columbia physicist I. I. Rabi agreed with Bridgman: "The scientist cannot take responsibility for the manner in which society utilizes the knowledge he uncovers" (1948, p. 73).

Occasionally pleas were voiced like Bertrand Russell's (1960, p. 391):

> The scientist is also a citizen; and citizens who have any special skill have a public duty to see, as far as they can, that their skill is utilized in accordance with the public interest. . . . It is impossible in the modern world for a man of science to say with any honesty, "My business is to provide knowledge, and what use is made of the knowledge is not my responsibility."

Shortly after that, though, Edward Teller dismissed such suggestions, saying: "Every citizen, whether he is a politician or a farmer, a businessman or a scientist, has to carry his share of the greater responsibility that comes with greater power over nature. But a scientist has done his job as a scientist when that power has been demonstrated" (1962, p. 56).

It is hard to know the full views of Oppenheimer, Bridgman, Rabi, and Teller on these matters. The record shows that during a war in whose struggle they believed, all had felt a responsibility to apply their scientific talents vigorously. All served in high R&D posts during the war. Later they advised government on atomic and hydrogen bomb issues and played important roles as educators and leaders. Perhaps they just defined the problem away by calling themselves scientists only when performing shirtsleeves labor at the blackboard or bench, and not when testifying before congressional committees on the Superbomb, politicking for the federal physics research budget, or advising on weapons-testing protocols. Perhaps they preferred to have nonscientists believe that science is done in a social vacuum by absentminded, apolitical souls who, like poets, stroll through intellectual gardens waiting for ideas to drift into their heads. But they all knew that solving complex "dirty" practical problems requires that very abstract ideas be solicited, cultivated, and sweated over, often by teams of experts, before they can even begin to be applied. And they all knew that no project as ambitious as the H-bomb could be achieved without heavy campaigning—for political support, money, facilities, and talent—by highly accomplished, publicly recognized scientists. Confusion was engendered in part by depending on the ambiguous word "responsibility" and in part by focusing on scientists as individuals rather than as members of groups.

The 1960s and 1970s brought some sensitization and sorting-out, although most of the commentary was not very incisive. Too, activities vaguely construed as science or technology were blamed variously, and often wrongly, for an inhumane Vietnam War, growth of tacky suburbs, and increased cancer mortality.

Gradually, with the publication of Don Price's influential book, *The Scientific Estate* (1965), and discussions in the American Association for the Advancement of Science and other forums, the nature of the tacit social compact surrounding scientific work became clearer.[17] But despite repeated implorations for responsible action, to this day the notion remains vague and uncompelling. Thus it is not surprising to hear Samuel Florman, a prominent spokesman for the engineering profession, complain: "For those of us who are in search of a new philosophy of engineering, the new morality of 'social responsibility' appears, upon inspection, to be pretty shallow stuff" (1976, p. 31).

Semantic dilution of the term "responsibility," from its carrying several somewhat conflicting common meanings, greatly weakens the concept's power. "Responsible" may connote causal agency: "His philandering was responsible for their divorce." Or authorized duty: "The police are responsible for controlling traffic." Or reliability: "That is a responsible analytic laboratory." Or capability of motivation: "Was the accused psychologically responsible at the moment of the crime?" Or ability to respond as charged: "The FDA is responsible to congressional direction." Or ethicality in conduct: "She's a responsible journalist." Or holdability to account: "By law the corporation is responsible for protecting its workers from injury."[18]

Thus it may be puzzling but is not implausible to say, in effect, "Yes, he is responsible [for some action], but he can't be held responsible"—that is, "He had a part in it, but no specific compact exists by which he can be judged derelict or held liable for reparation."

Only in some binding context of trust can people's actions be held up to evaluative criteria. A physician must be responsible to standards of prevailing medical practice, common and tort law, and his employment contract. A molecular biologist whose research is supported by federal funds must respond to federal guidelines governing laboratory practice and protection of human experimental subjects, to rules established by his institution and by local, state, and federal governments, and to collegial expectations that he will exercise due care.

17. American Association for the Advancement of Science, Committee on Science in the Promotion of Human Welfare, 1965, and Committee on Scientific Freedom and Responsibility, 1975; and Edsall, 1975.

18. For similar distinctions, see Hart, 1968, pp. 211–230; Haydon, 1978; and Jonas, 1984.

The really troublesome situations are those in which the science or technology are novel and of uncertain effect, those about which moral consensus and expectations have not been clarified, and those for which evaluative social guides (rules, treaties, laws, ethical codes) have not been developed. Calling for individuals in such cases to "be responsible" simply doesn't carry much compunction to take initiative, change ways, feel regret or guilt, or make reparation.

To conceive of a person's "acting purely as a scientist" begs many issues. As I am arguing all along, technical action is deeply—consciously and subconsciously—influenced by personal values and by auspices of work. A scientist is specifically a Canadian Department of Fisheries and Oceans biologist assessing the effects of acid rain, a Hanford Laboratory chemist studying the properties of weapons-grade plutonium, or a Teamsters Union medical advisor counseling on lower-back injuries—and a mother, a Catholic, a Greenpeace supporter, a technological optimist, a hawk, a diabetic, a Republican, a mathematical inadept, a proabortionist, a Greek. . . .

Similarly, to hail "loyalty to science" is, for most purposes, to swear allegiance to an outmoded abstraction. The only sense in which such loyalty can be binding is when it means telling the scientific truth as understood and proceeding by the internal quality-control norms of one's particular scientific community.

In most technical areas the one-to-one personal relationships that classically governed ethical conduct have been supplanted by much more diffuse ones buffered by intermediaries. Where once a bridge builder whose bridge collapsed might have had to run from an angry mob, usually now he faces nothing worse than—although some would say the punishment is equivalent—serving time in lawyers' offices and in court. Engineers who design roads, tunnels, bridges, dams, and airports interact with their ultimate clients, the public, only through managers, attorneys, and the officials who supervise public contracts. Vaccine developers plan their research in committees. Many applied-social-science researchers never even meet the people their studies examine and affect. Physicians still carry the wand of Aesculapius, but they do so in the context of the giant medical-care industry.

Diffuseness of responsibility has resulted in part from extraordinary enlargement of our social ambitions. Health has long been a subject of social and professional concern; but today's national health-care policies and programs, which define health broadly and intend to cover all citizens, have valuative dimensions that simple private-physician incentives and classical medical ethics can hardly be expected to accommodate.

Diffuseness has also resulted from growth in the dynamic complexity of technological undertakings. I can cite no clearer example than two simultaneous accounts of the National Transportation Safety Board's investigation into the cause of the DC-10 wreck in Chicago in 1979: the *New York*

Times headlined, "Everyone Blamed in DC-10 Crash"; the *San Francisco Chronicle*, "No One to Blame for DC-10 Disaster." Given the hundreds of technical experts and officials in the McDonnell Douglas Corporation who designed, built, and established maintenance procedures for the airplane, and the hundreds more in American Airlines who operated, maintained, and inspected it, and the hundreds more in the Federal Aviation Administration who regulated all these matters, it is not surprising to have to conclude that everybody involved was responsible but none could be held responsible.[19] Culpability for the Three Mile Island accident was laid not simply on the operators, or the builders, or the electric utility owners, or the Nuclear Regulatory Commission, but, in a strong sense, on the entire nuclear power establishment—which has responded collectively by taking such actions as setting up the Nuclear Safety Analysis Center to analyze operational problems on an industrywide basis.

The possibilities and frustrations in trying to be responsible are illustrated by the charter of the new organization, Computer Professionals for Social Responsibility, "an alliance of computer professionals who are alarmed about the growing catastrophic threat of nuclear war." These experts, many of whom work on the computers that control America's nuclear defense systems, in 1984 resolved:[20]

- We will encourage wide recognition of the global threat posed by the nuclear arms race.
- We will attempt to understand better the role that we and our technology play in the militarization of our nation and the world.
- Within the computer science community, we will encourage debate and reexamination of the military uses of our technology.
- We will explore the ways in which general confidence in the expertise of our community has discouraged public debate of, and individual responsibility for, our national defense policies.
- We will work to dispel popular myths about the infallibility of technological systems, and about the adequacy of technological solutions to our security problems.
- We will work with scientists and professionals in other countries to build mutual understanding, and will urge our respective governments to abandon the threat or use of force as a method of resolving conflict.
- We will promote the use of our skills in combatting the causes of war and in serving constructive human purposes.

19. Newhouse, 1982, traces general issues of design philosophy and management surrounding the DC-10.

20. Computer Professionals for Social Responsibility, P.O. Box 717, Palo Alto, California 94301.

It would be futile to expect individuals to respond alike to these problems. While some researchers never like to leave the laboratory, others find it engaging and rewarding to serve on committees, hold office, or shuttle to Washington, Geneva, or Nairobi to give advice. In the United States, moral views may vary even within families, and certainly the country does not fully embrace a shared morality beyond some grand ambiguous proclamations and more specific but weakly articulated and inconsistent intentions. Oughts are far from clear. Experts are as likely to take courageous initiatives because they love the threatened Maine coastland, or because as parents they are worried about asbestos in a school, or because their own research activity is implicated, as they are because they feel a call of extraordinary duty on holding a technical degree and being employed at some technical work.

Arthur Galston, recalling the way his early research on plant maturation factors contributed to the eventual industrial development and military use of Agent Orange, described the complications of technical parentage (1972, p. 233):

> I used to think that one could avoid involvement in the antisocial consequences of science simply by not working on any project that might be turned to evil or destructive ends. I have learned that things are not all that simple, and that almost any scientific finding can be perverted or twisted under appropriate societal pressures. In my view, the only recourse for a scientist concerned about the social consequences of his work is to remain involved with it to the end. His responsibility to society does not cease with publication of a definitive scientific paper. Rather, if his discovery is translated into some impact on the world outside the laboratory, he will, in most instances, want to follow through to see that it is used for constructive rather than anti-human purposes. But I know of no moral imperative to invoke here; some individuals feel moved to respond to the social challenge, while others shun such activity, either through timidity, aversion to political argumentation or a feeling that others, better trained, should handle social problems.

Scientists indeed have chosen variously. Recounting that in developing napalm during World War II he was "working on a technical problem that was considered pressing," Harvard chemist Louis Fieser said, "I distinguish between developing a munition of some kind and using it. . . . I'd do it again, if called upon, in defense of the country."[21] On the contrary, during the Vietnam War MIT physicist Steven Weinberg and some other scientists refused to advise the Defense Department (Steven Weinberg, 1974, p. 35).

---

21. "Napalm Inventor Discounts 'Guilt,'" *New York Times*, December 27, 1967.

Robert Wilson, who directed the experimental physics division of Los Alamos during the atomic bomb project, later described the in-or-out dilemma associated especially with national security issues (1970, p. 33):

> As soon as possible, I returned to university life, renouncing anything further to do with weapons work, or, in fact, with any kind of work connected with secrecy. But even that kind of holier-than-thou course of action has since caused me considerable qualms. It was a kind of cop-out that is all too manifest in our youth today: good for my conscience, perhaps, but it immediately reduced my effectiveness to do something about nuclear energy. My expertise soon became outdated: I had to watch my more conservative friends, usually working from within the government, give the kind of advice and exert the kind of political pressure that is based upon understanding.

Wherever contractual or traditional expectations are clearly agreed on, responsible action by practitioners obviously should be expected and enforced. But, especially for important complex problems, nonspecific responsibility can neither sufficiently motivate nor provide guidance for individual scientists' actions. Even extraordinary efforts by extraordinary individuals will be inadequate against some of the novel and diffuse problems we face today. Collective approaches must be sought that do not depend on heroic initiative or sacrifice. These approaches must be based not so much on legalistic compulsion as on inducement and stewardship.

Five

The Architectonics
of Technical Trust

If the principles and admonishments of professional responsibility are so ineffective in addressing large diffuse social problems, if applied scientific work is so judgment-bound, if the constraints that apply to professionals like dentists apply so weakly to people like molecular biologists who affect our lives powerfully though indirectly, what *structural* ethical recourses are available?

Foremost—more important than any grand reform—we need to recognize what actually does and can happen within existing structures, and to upgrade those structures to meet the new kinds of problems.

A triad of interlocking elements comprise the essential architectonics of technical trust: technical communities are the core operative units; bargainable compacts between those technical communities and society are the dynamic frameworks; and peer review is the crucial quality assurance mechanism.

I use the heavy word "architectonics" because what is at issue is deep structure, much of it implicit rather than explicit, that transcends conventionally described organizations and procedures.

The key, I am convinced, to making sense of the whole modern technical enterprise and "getting handles" on its diverse manifestations is to recognize how all aspects of research, development, and technical public service are governed by bargainable compacts—compacts between, on the one side, those who pursue science and apply technology, and, on the other, those who pay for it, are affected by it, and expect, or at least hope for, its service.

Construing this activity as compactual has the advantage that it couples technical communities' internal mores and quality assurance to their external social stewardship. It couples freedoms to constraints. It resituates professionalism. It reveals pattern in seemingly disparate sociotechnical controversies. It applies in all countries and cultures. And it suggests avenues of remedy.

TECHNICAL COMMUNITIES, THE CORE OPERATIVE UNITS

One of the central developments in modern science and technology has been the coalescence of operative technical communities—not merely groups of people sharing common interests, but organized, self-reflective groups of specialists who pursue communal goals. They vigorously communicate and validate ideas, findings, and methods, exchange specimens and colleagues, promote their fields, and render public service.[1]

Although reference often is made to "the scientific community," this, like "the artistic community" or "the business community," comprises such a large amorphous population of so many diverse functional subcommunities that almost always it is more valid to speak of particular scientific and technical communit*ies*.

Technical experts are quite specialized and compartmentalized. Nuclear engineers cannot assess the genetic effects of radiation, and radiation biologists cannot analyze in detail the nuclear chemistry of decaying fuel elements. Not only do botanists not read astronomy journals, lichenologists are not likely to follow the botany of grasses. I doubt that any chemist alive would consider himself qualified to referee for publication the diversity of manuscripts that comprise one issue of the *Journal of Organic Chemistry*, much less those for the broader *Journal of the American Chemical Society*.

The principal activity American scientists share in common is scanning the news and gossip sections of *Science* magazine—and engineers, doctors, and some other applied scientists mainly read their own journals. The American Association for the Advancement of Science, a society of some 138,000 members (which is, by any count, only a modest portion of the national scientific and technical population), provides numerous coordinative services and plays important leadership roles, but it is unwieldy both as facilitator and as lobby, and it has few ways of canvassing or expressing its members' consensus.[2]

For most purposes, more tightly defined groups have to be viewed as the operative communities. They may be problem-oriented (diabetes researchers) or discipline-oriented (crystallographers). Usually they function through formal organizations, such as professional societies, elected academies, and trade or special-interest associations. Technical communities

1. Price, 1965, Hagstrom, 1965, and Crane, 1972, discuss aspects of scientific community.

2. For history of the AAAS's early role in focusing the American scientific endeavor, see Kohlstedt, 1976. Also see MacLeod and Collins, 1981, on the British Association for the Advancement of Science 1831–1981.

are not static but are constantly splitting, merging, and evolving. As the sociobiology, recombinant DNA, and other examples in this book illustrate, they also cross-critique each other's work.

In many fields no single professional organization can fairly be taken to be identical with, and completely represent, the community. Despite newsmedia intimations to the effect, the American Medical Association is not synonomous with the entirety of the American medical community.

To conceive how an R&D or professional community "works" one usually must define it as including not only primary practitioners but also administrators of leading institutions, key bureaucrats (such as federal budget officers), relevant technical societies, and the cluster of businesses, service firms, publishers, and communication, education, and certification organizations that constitute its infrastructure.

Contrary to the stereotypical image of scientists as antisocial loners, technical people operate by committee more than any other professions do. Literally thousands of organizations and committees work to organize practitioners in general (the Geochemical Society), to set standards for professional practice (American College of Obstetricians and Gynecologists' Committee on Professional Standards), to recognize accomplishment (the Association of Official Analytical Chemists' committee to grant the Harvey W. Wiley Award for the Development of Analytical Methods), to advise on R&D priorities (High Energy Physics Advisory Panel to the Department of Energy), to referee research proposals (National Institutes of Health "Study Sections"), to direct the overall work of institutions (Board of Trustees of the Woods Hole Oceanographic Institution), to referee and oversee publications (Editorial Board of the *Journal of Cell Biology*), to tend infrastructure resources (American Mathematical Society's Committee on Translations from Chinese), to develop nomenclature (nomenclature committees of the International Union of Pure and Applied Chemistry), to oversee education and certification (Registry Committee of the National Registry of Microbiologists), to attend to personnel issues (American Chemical Society Committee on Women Chemists), to advise on practical problems (Transportation Research Board Committee on Batching, Mixing, Placing, and Curing of Concrete), to advise on technical policy issues (the President's Cancer Panel), to develop and promote special-interest views (Atomic Industrial Forum), and to organize collaborative research and exchange (Committee on Scholarly Communication with the People's Republic of China).

If we look for a centerpiece of the American technical enterprise, we find one of the most vigorous and prestigious establishments to be the cluster of National Academy of Sciences institutions. This is as good a place in the book as any to comment on it. The parent Academy, a self-perpetuating honorific society of some 1,300 elected members, is generally considered the elite body of American science, although like all academies it sometimes fails to elect some accomplished experts. Affiliated with it are the 1,000-member National Academy of Engineering and the 400-member Institute of Medicine. These three academies jointly oversee the National Research Council, a floating talent bank of about 1,000 committees, subcommittees, boards, and panels comprising more than 9,000 experts from all over the nation. NRC committee members serve without compensation and prepare reports on an enormous variety of issues. The Academy and its affiliates enjoy quasi-governmental tax-free status, like the American Red Cross, are not registered as lobbying organizations, and have pleaded legal immunity from the Public Advisory Committee Act and the Public Information Act. Whereas traditionally the Academy at least maintained a facade of rendering dispassionate technical advice when called upon, now it vigorously campaigns to get itself commissioned to perform studies, follows many of its reports "out into the world" to make sure they receive attention, and takes stands on quite political issues. Repeatedly the Academy has to insist to journalists and the Congress that it is *not* a government agency. Lately the Academy leadership has disavowed support for some NRC reports, and has seemed to distance itself from others. The title page on NRC reports now omits Academy credit, naming only the committee involved, and the Institute of Medicine cites its reports by names of committee-chairman and staff-director "editors." Although I wholeheartedly admire most of the Academy's public service endeavors, I fear that the body is taking on too many issues and diluting its usefulness, is declining to take full institutional responsibility for many of its studies, and is courting further politicization. If NRC committees become indistinguishable from ordinary technical advisory committees, the Research Council might as well be converted into a permanently staffed government agency to convene ad hoc committees, with members nominated by the elected Academy, to provide assessment and advice for other agencies. The independence the Academy preserved, even if precariously, has served us extraordinarily well, and we should hope the Academy's intellectual authority will be tended carefully.

BARGAINABLE COMPACTS, THE DYNAMIC FRAMEWORKS

Technical activity is governed by both explicit and tacit compacts with society. Each community works out its own arrangements.

Society *invests in* the training, professional development, and general work of technical communities. It invests heavily: including research facilities and instruments, information banks, communication systems, and other aspects of infrastructure, as well as R&D grants and contracts, substantial public subsidy of one form or another goes to virtually every college, university, medical center, field station, and research facility in the United States. Few technical people can claim not to have benefited from this largesse. The situation is similar in most other countries. For the most part the technical professions are left free to govern themselves, control admission to membership, direct their own research, enforce quality of work, and advise on allocation of public and semipublic funds.

Concomitantly, society *invests with* the technical professions and their institutions certain trusts. This tacit compact gives rise to ethical oughts. In return for granting of status, support, and freedom, the public holds the technical community to a variety of expectations. Despite their informality, such compacts are crucial to our diversely specialized society.

Technical practitioners develop two kinds of obligations. The first is an ethics of maintenance of the profession, based on internal standards and restraints mutually agreed to by peers, and shading into being an etiquette: publishing experimental details fastidiously, refraining from pilfering ideas and making false claims, adhering to standards of service and contract, contributing to the training of protégés.

The second obligation is a commitment to the service of society as well as to individual clients. This commitment usually grows slowly. By stepping into gaps during crises, or by deliberately staking out pieces of social territory, professions come to be responsible for special matters. Not only have geneticists dispassionately studied our chromosomes as objects of scientific curiosity, but by subtle stages they, along with related specialists, have also become guardians of our genetic legacy. In part this guardianship is self-appointed. It is rarely contested. Because this particular compact seems to work—that is, geneticists earn support and freedom, and society secures a watch over its gene pool—society now tends to expect geneticists, as a matter of responsibility, to stand watch for mutagenic menaces. In general, geneticists seem to accept the responsibility, both as individuals and through organizations such as the Environmental Mutagen Society, The American Society of Human Genetics, the March of Dimes Birth Defects Foundation, committees of the National Research Council, and the International Council on Radiological Protection.

Such expectations are not unique to science. The goldsmiths' trade is predicated on integrity of the fineness-mark—14-karat gold has to be guaranteed to contain 14 karats, otherwise the whole enterprise would collapse. Despite abuses by some lawyers, by and large we must count on the legal profession to supervise both the law's philosophical evolution and its daily practice. It is hard to imagine otherwise. By the very nature of things, state-of-art skill and effective entrée must be vested in recognized expert gold assayers and lawyers. So with the technical professions.

Professionalism-as-compact embraces change by allowing that from time to time compacts can be modified. Someone, perhaps an outsider, perhaps a member of the profession, levels a charge of irresponsibility or corruption. In response the profession purges itself of charlatans or revises its code of practice. Or the public withdraws support or imposes licensing requirements. The courts try accusations of medical or engineering malpractice. Clean-house-or-we'll-clean-it-for-you challenges ran mesmerists, nostrum-flacks, hack abortionists, bleeders, and purgers out of the medical profession in the early years of this century. Recently similar pressures have induced scrutiny of psychiatrists suspected of taking unethical sexual advantage of clients, and of industrial scientists alleged to have suppressed toxicological data. Displeased with the way the Three Mile Island reactor operators handled their 1979 emergency, society, via government regulators, the Congress, and electric utility directors, is demanding higher standards for reactor personnel selection, training, and licensing.

New or renewed activities seek to be included within existing compacts or to negotiate their own. Currently midwives are seeking both, as they fight for acceptance from obstetricians and anesthesiologists and hospitals, and as they demand state licensing and eligibility for reimbursement under health insurance schemes. Acupuncture, sexology, hypnosis, and various schools of nutritional therapy and holistic medicine—all claiming scientific soundness—similarly are struggling for legitimate status.[3]

Compactual vigilance covers basic research as well as direct client–professional service. Typical of formulations is the warning on biomedical research fraud issued by the Association of American Medical Colleges (1982):

> The maintenance of public trust in this pursuit is vital to the continuing vigor of the biomedical research enterprise. Loss of this trust because of isolated instances of dishonest behavior on the part of a few researchers could cause great harm by calling into question in the mind of the public the validity of all new knowledge and the integrity of the scientific community at large. In short, it is in the best interest of the public and of academic medicine to prevent misconduct in research and to deal effectively and responsibly with instances where misconduct is suspected.

The professions not only face inward and enforce their codes, but also outward to protect members who meet resistance in discharging their obligations. On occasion an employer may interpret a professional's attempts at responsible action as disloyalty or worse, and harass or threaten to fire or sue. The repressed professional may retreat to the sanctuary of the guildhall. In instances I will describe later, professional societies have provided legal defense and supportive publicity for members persecuted unduly for "blowing the whistle" on what they believed were unconscionable situations. Many professional societies now have ethics committees to review cases and to prepare for whistleblowing contingencies. An organization may choose not to defend its members' opinions, but rather to defend their right to raise warnings without recrimination. An association

3. McRae, 1982, reviews regulation of acupuncture.

may advise its members of ways to exert influence reasonably, and it may offer help in resolving accusatory confrontations.

The challenge to professionalism in fulfilling society's needs—including, sometimes, even needs society is not yet aware of—can be stated in two questions. Within its compact as currently interpreted, does the profession meet expectations? And beyond that, does it fill the more general needs of society that hinge on expertise presided over by, or on authority vested in, the profession, but which may not be precisely specifiable as one-to-one provider–client obligations?

Physicians obviously perform masterfully in standard therapeutic practice. The difficulties lie in the scope of "standard practice." American physicians do not devote as much attention to preventive medicine or health promotion as I think they could. On the two dominant fatal illnesses, cancer and heart disease, doctors traditionally have tried little to affect such reducible causes as occupational exposure to chemicals, tobacco smoking, and the overweight–underexercise syndrome. Venereal disease, sensory deterioration, and other conditions similarly have been given relatively little attention.

This is the inherent limitation of professionalism: its power extends mainly over defined, accepted practice. Operational limitations have to do with whether the profession actually does enforce standards. In medicine, for example, the criteria and mechanisms for censure are firmly in place, and they are applied to an extent. At question is the aggressiveness and integrity with which malpractice actually is curtailed.

Over the past two decades applied scientists, such as toxicologists and analytic chemists, have been reexamining their social roles. Stimulus has come both from public dissatisfaction and from friction with intersecting professions. Most reforms have addressed internal ethics, and ethical relations with identifiable clients, rather than broad issues of public service. There is little evidence that the majority of members of professional societies, who join mainly to be able to list the affiliation on their résumés, attend conferences, and receive the journals, have been affected.

In recent years most technical fields have had to renegotiate their compacts with society. Anatomy and taxonomy have had to defend their usefulness. A coalition of disparate practitioners has sought recognition of a hybrid new field, materials science, as has the coalition working on artificial intelligence. Psychology has had to restrain overzealous researchers who violated subjects' rights. Biology education has been challenged by proponents of biblical creationism. Biomedical sciences have had to castigate instances of research fraud. Consulting toxicology laboratories, confronted with several cases of testing fraud in the industry, are facing proposals that practitioners be licensed and laboratories certified. Doctors have been sued by chiropractors for monopolism. All of the sciences have had to compete for financial support.

PEER REVIEW, THE QUALITY ASSURANCE MECHANISM

Science grows through cumulative conjecture, discovery, criticism, and rejection or endorsement of ideas by the dispersed body of the world's scientists. Institutionalized skepticism in the form of peer review, and perennial willingness to submit research to review and challenge, are the mechanism and attitude that make science more—so much more—than just a collection of personal opinions.

A major consequence of science's increasing involvement in the public agenda has been that, rather than simply ignoring weak science, now scientists are being asked to devote much effort to critiquing it and, where deserved, actively discrediting it.

Communal authority may be deferred to as it was by Margery Shaw in a letter to the editor of *Science*, concerning controversial studies of chromosome damage in Love Canal residents (1980):

> I suggest that the cytogenetics community attempt to design a study that would be acceptable *in advance*, considering all of the parameters, such as culture conditions, intraobserver consistency, intraobserver differences, suitable control groups, appropriate staining procedures, number of cells per individual and number of individuals to be scored, number of laboratories, and blind scoring of subjects.

Similarly in the social sciences. Noting that "many surveys are not reviewed at all within the scientific community" because of "the applied nature of survey research and the institutional structure of the industry," the 1981 National Research Council report on *Surveys of Subjective Phenomena* recommended that peer review be made a part of the design, analysis, and reporting of surveys.[4]

Government fact-finders may seek technical communal certification of facts. Needing interpretation of lung and liver sections from four laboratories' mouse and rat tests of the pesticide heptachlor, the chief administrative law judge of the Environmental Protection Agency asked for an audit by the Pesticide Information Review and Evaluation Committee of the National Research Council. Five duly appointed academic pathologists examined 2,753 slides, labels masked, through a five-headed microscope and scored the specimens as being normal, precancerous, or cancerous. This provided the judge with authoritative information for his heptachlor registration proceedings.[5]

Peer review should not be expected in and of itself to transmute the fundamentally adversarial nature of public disputes. Science is, in its essence, argumentative. Disagreements still will have to be fought out.

---

4. NRC, Panel on Survey Measurement of Subjective Phenomena, 1981, p. 52.

5. NAS/NRC, Pesticide Information Review and Evaluation Committee, 1977.

Peer review can filter out sloppy science, or partisan "loaded" science, and can identify oversights, omissions, and inconsistencies. By removing "red herring" distractions and defining the sphere of disputes it can help reduce acrimony. Along with other procedures it can promote fair, good-faith exchange. Nonetheless, technical staff and advisors of the various interests will play their institutional roles; because of selection and self-selection for roles, personal, disciplinary, and institutional biases will remain partly determinative.

Possibilities for conflicts of interest abound. Highly regarded scientists shuttle between industry, universities, the government, and other employers throughout their careers. A Nuclear Regulatory Commission engineer may formerly have worked for a reactor manufacturer, or hope to work for one in the future. An industrial engineer may hold dues-paying membership in the Sierra Club engaged in a lawsuit against his firm.

The term "peer review" is used much too cavalierly. Usually what really is being called for is elite review, criticism not just by people who do passable work, but by people who have earned laurels of high scientific recognition. Not just everyone employed as a scientist is a peer—or at least there are several levels of peerage. Thus when a special panel of the Nuclear Regulatory Commission critiqued the Rasmussen *Reactor Safety Study* it complained (U.S. Nuclear Regulatory Commission, 1978, p. 2):

> Effective peer review is the only method known to the technical community for quality assurance in its product. However, in the arena of reactor safety, a peer comment has come to mean anything written by anybody to anybody asserting anything about anything. The comment need not contain a better analysis, demonstrated technical expertise in the subject, evidence of error, or conformance to any of the normal standards by which the technical community assesses a peer comment. Thus, responsible peer comments, which *do* exist, are immersed in a sea of others, and this degrades the peer review process.

In the intellectual marketplace some people develop reputations as perceptive, fair, constructive reviewers, while others become known as narrowly focused, inflexible, or unfair. Selection is merit-based. Admonished by lawyer Peter Hutt that Academy committee members should be selected by a more democratic process and that committees should follow "open and proper procedure" in their government advisory work, National Academy of Sciences president Philip Handler responded (*Science* 191, p. 543, 1976):

> We choose the members of our committees with extreme care. We have no sense of participatory democracy. This *is* an elitist organization, sir. We go to great care to elect the members of the Academy and we are guided by their insights. To have a democratic process by

which the committee is then brought into being is to give away the only special asset we have in this building.

A constructive development of recent years has been professional groups' increased willingness to review major assessments. In 1954 the American Statistical Association prepared a detailed review, *Statistical Problems of the Kinsey Report*, of that study's overall design, sampling and interview procedures, analytical methods, recordkeeping, presentation of data, and interpretation (Cochran, Mosteller, and Tukey, 1954). In the late 1970s when the National Center for Toxicological Research, the federal agency, undertook its elaborate 24,192-mouse one-percent-effective-carcinogen-dose (ED_{01}) study, several professional groups, mainly the Society of Toxicology, critiqued the experiment's design before, during, and after the laboratory research was carried out (Society of Toxicology, ED_{01} Task Force, 1981). Over the decade 1972–1982, at the request of the U.S. Food and Drug Administration a select committee of the Federation of American Societies for Experimental Biology evaluated voluminous toxicological data on 468 food ingredients, such as black pepper, pectin, yellow mustard, benzoic acid, and clove oil, that long have been "generally regarded as safe."[6]

Applied mathematicians are of course essential in most assessments, for help in designing tests, auditing calculations, and assisting in the drawing of conclusions. Much of this is routine craftwork. But one unusual, crucial role should be recognized. It is my guess that statisticians Frederick Mosteller (Harvard) and John Tukey (Princeton) have served on or assisted more technical committees than anybody else alive, the principal competitors for this honor being their fellow mathematicians Saunders MacLane (Chicago) and Isadore Singer (Berkeley), who for years stalwartly led the National Academy of Sciences' internal draft-report review process. It is not just their ability to manipulate numbers that keeps these experts in demand, but sensibility in thinking through questions of macro-experimental design: how inquiries should be cast, what evidence and logic are applicable, how discrimination can be increased, how uncertainties and sensitivities should be probed, what inferences are allowable from evidence. Mosteller and Tukey outlined this role in an article in 1949 in which they called for education of "scientific generalists" who would master "science, not sciences" (Bode et al., 1949). Probably we cannot train people to become such generalists, but we can try to cultivate talent when it emerges.

From time to time expert communities may need to engage in collective self-review and critique the use of their techniques. For example, concerned about the validity of some methods and about standardization for

6. Select Committee on GRAS Substances, 1982, describes the auspices and procedure of the review.

interlaboratory comparison, two American Chemical Society committees have developed guidelines for environmental chemical analysis.[7] Psychiatrists have performed field trials to assess the reliability with which different therapists using the *Diagnostic and Statistical Manual of Mental Disorders* arrive at the same diagnosis of disorders (Spritzer, Forman, and Nee, 1979). And the American Joint Committee on Cancer, which involves several major biomedical societies, convened nineteen anatomic-region task forces to develop detailed criteria for classifying cancers (1983).

NEGOTIABLE FACTORS

When, at whatever initiative, technical endeavors adjust their place and roles in society, the following factors in the technical community–peer review–dynamic compact architectonics can come up for negotiation. All will be discussed in more detail in later chapters.

SELF-DETERMINATION. Discoveries take time, support, and room. As with all creative work, flexibility is essential. Bureaucratic harassment, short-term constraints, demands for precise plans, and interference with reporting of results are seriously debilitating. Nonexperts can provide guidance, but nobody other than the specialists involved, and perhaps experts from intersecting disciplines, can fully comprehend the direction and promise of a field. Basic researchers, especially, fight for self-determination.

SUPPORT. Technical work obviously depends on money and personnel, and often on unusual forms of support, such as access to rare sites or phenomena. More than many endeavors it has to maintain a robust infrastructure of recruitment, training, instrumentation, information systems, specimen collections, and special support skills such as glassblowing, instrument making, and experimental animal husbandry. As in other risktaking ventures, vital overhead must be paid to keep the basic intellectual fund of research thriving, both to seed new growth and to be drawn upon for application. A few years of neglect can seriously erode these resources. Society holds the purse strings.

GENERAL HEALTH OF THE TECHNICAL COMMUNITY. Periodically, technical groups need to take stock and revitalize. At various times in the past two decades committees of the National Academy complex, jointly with related organizations, have surveyed the prospects, opportunities, and policy issues in their research areas, to considerable benefit (Lowrance, 1977).

---

7. American Chemical Society, Committee on Environmental Improvement and Subcommittee on Environmental Analytical Chemistry, 1980.

RESEARCH AND DEVELOPMENT AGENDA. Technical and nontechnical leaders have to work together to set priorities, allocate resources, decide who is going to do what, and determine the conditions and timing under which work will proceed. In established fields this usually is hard-fought but clearly structured; in emerging areas, structuring the agenda process itself is part of the problem.

QUALITY ASSURANCE, IN PUBLIC CONTEXT. Fields must enforce quality in research and application. Naturally, this serves the fields' self-interests. But beyond that, technical claims need to be filtered for public decisions. This requires preventing premature release of findings, criticizing too-narrow construal of sociotechnical problems, censuring technical overclaiming and overpromising, sorting out fact/value hybrid problems, and interpreting issues to the lay public and its leaders.

RANGE OF APPLICATION. Fields such as epidemiology and criminology increasingly are being pressed to expand their reach of application, to put their theory at the service of social problems not dealt with before (epidemiology in search of behavioral illnesses, for example, and criminology, white-collar crime). The challenge is for them to respond without overreaching.

ANTICIPATION OF SOCIOTECHNICAL CHANGE. More than ever there is a need to assess implications of technical change in advance, and to trace consequences. Anticipatory approaches must be developed. Political and scientific leaders need to improve ways for encouraging the raising of concerns, and for setting exploratory agendas.

PARTICIPATION IN SOCIETY'S DECISIONS OF CONSCIENCE. Technical experts can greatly contribute to the defining and sorting out of decisions and help adjust the timing, pace, and complexion of technical advance. On most issues the "middle" range of technical and social opinion needs to be represented much more fully.

A VIVID CASE OF COMPACT-RENEGOTIATION: RECOMBINANT DNA RESEARCH

> We are writing to [the presidents of the National Academy of Sciences and the Institute of Medicine], on behalf of a number of scientists, to communicate a matter of deep concern. . . . We presently have the technical ability to join together, covalently, DNA molecules from diverse sources. . . . Certain such hybrid molecules may prove hazardous to laboratory workers and to the public. Although no hazard has yet been established, prudence

suggests that the potential hazard be seriously considered.—
Maxine Singer and Dieter Soll, senior research scientists,
National Institutes of Health (letter to the editor, *Science* 181, p.
1114, 1973)

At present, the hazards may be guessed at, speculated about, or
voted upon, but they cannot be known absolutely in the absence
of firm experimental data—and, unfortunately, the needed data
were, more often than not, unavailable.—*NIH Guidelines for
Research Involving Recombinant DNA Molecules* (41 *Federal Register*, p. 27,911, 1976)

I set my hope in the cleaning women and the animal attendants
employed in laboratories playing games with "recombinant
DNA"; in the law profession, which ought to recognize a golden
opportunity for biological malpractice suits; and in the juries
that dislike all forms of doctors.—Erwin Chargaff, emeritus professor, Columbia University (letter to the editor, *Science* 192, p.
938, 1976)

DNA splicing research, far from being an idle scientific toy or
the basis for expensive and specialized aid for the privileged few,
promises some of the most pervasive benefits for the public
health since the discovery and promulgation of antibiotics. . . .
And perhaps the most important products are those that remain
to be discovered.—Joshua Lederberg, professor, Stanford University ([American Medical Association] *Prism* 3, no. 10, pp. 37
and 36, 1975)

The real hazard is the one no one . . . has dreamed of yet, and
this you cannot specify against.—DeWitt Stetten, deputy director, National Institutes of Health (quoted by Francine Simring
in a letter to the editor, *Science* 192, p. 940, 1976)

I think [the Guidelines] are totally capricious and totally unnecessary.—James Watson, director, Cold Spring Harbor Laboratory (*CoEvolution Quarterly* 14, p. 40, Summer 1977)

If right now I had to weigh the probabilities of either event I
would guess that recombinant DNA research carries more and
earlier risks of causing cancers than hope of curing them.—
George Wald, professor, Harvard University (*The Sciences* 16, no.
5, p. 10, 1976)

If today's imaginative rhetoric about the dangers of recombinant
DNA research had been in fashion 50 years ago, voices would
have been raised against the use of staphylococci or poliomyelitis
virus in laboratories, and we might have lost the information

> which led, ultimately and quite unpredictably, to penicillin and the polio vaccine. . . . I am not so much in favor of recombinant DNA research as I am opposed to the opposition to this line of inquiry.—Lewis Thomas, president, Memorial Sloan–Kettering Cancer Center (testimony before the U.S. House of Representatives, Subcommittee on Science, Research and Technology, 1977)[8]

> It is possible that the "recombinant DNA affair" will someday be regarded as a social aberration, with the Guidelines preserved under glass. Even so, we can say the beginnings were honorable. Faced with real questions of theoretical risks, the scientists paused and then decided to proceed with caution. That decision gave rise to dangerous overreaction and exploitation, which gravely obstructed the subsequent course. Uncertainty of risk, however, is a compelling reason for caution. It will occur again in some areas of scientific research, and the initial response must be the same.—Donald Fredrickson, director, National Institutes of Health (to a conference in 1979; reprinted in Morgan and Whelan, 1979, p. 399)

The controversy that swirled among these opinions in the mid-1970s was a serious test of scientists' stewardship. Altogether it amounted to a renegotiation of the social compact within which molecular biological R&D is pursued. It is in this light that I will review it here. In essence, a dream materialized before its dreamers and those affected were ready for it.

In the preceding decades, marvelously intertwining lines of research had demonstrated the fundamentals: that genes, the physical determinants of heritable characteristics, are segments of the long strands of DNA (deoxyribonucleic acid) that form chromosomes; that the way these genes dictate cellular activities is by coding for the structures of specific proteins, such as enzymes, which cells synthesize by transcribing from the DNA blueprints; that it is the particular sequence in which nucleic-acid subunits are arranged in a strand of DNA that determines the sequence in which amino-acid subunits become strung together into proteins, thereby determining the proteins' complexion and biological activity; that some genes serve to control other genes; and that the cellular labor of constructing, reading, and disassembling genes is performed by special enzymes, themselves proteins coded for by special genes.[9]

Geneticists surmised all along during this era of discovery, as Orwell and other fictionists less precisely had, that someday they would be able to alter

8. U.S. House of Representatives, Subcommittee on Science, Research and Technology, 1977, pp. 1151 and 1154; this testimony was based on Thomas, 1977.

9. Watson and Tooze, 1981, pp. 529–583, lucidly summarizes the scientific background of molecular genetics.

genes at the molecular level. In 1963 Rockefeller University molecular geneticist Edward Tatum implored a symposium (1965, p. 34):

> Before biological engineering is an accomplishment of the present rather than a possibility of the future, we must find time and energy to devote to the even more difficult question of how this knowledge can best be used for the welfare of all mankind. Perhaps we should begin to think seriously about this even now and to plan for the future in anticipation, rather than in retrospect, and possibly too late.

Salvador Luria, of MIT, warned that same symposium, "I fear the possibility that a negative genetic surgery may become available and that society—and geneticists themselves—may not be preadapted to cope with the dangers" (1965, p. 124). Listeners seem only to have politely agreed.

Although the genetic manipulation dream could have arrived from other sources, such as perhaps from cell fusion or chromosome transplantation experiments, it happened to come from gene mapping and the development of workaday molecular splicing tools. Key new tools were the restriction enzymes and related agents, isolated around 1970 from natural sources and made available as extraordinarily precise molecular scissors for clipping sequences out of DNA chains and coupling them to other chains. Other powerful new tools were plasmids, tiny primitive DNA molecules that live as internal parasites in bacteria and have an ability to convey acquired bits of foreign DNA, such as those that can be spliced into them by experimenters using the new coupling enzymes, between one host cell and another. Sometimes by clever laboratory contrivance the transplanted genes can be renaturalized in the new hosts, even inserted permanently into the hosts' own chromosomes, and be reproduced right along with the indigenous genes. Hence, "recombination."

For the first time ever, these discoveries made it possible not only to graft DNA but to repetitively reproduce (clone) genes and, beyond that, to cultivate foreign DNA in bacteria, yeast, or other easily grown species in large quantities and harvest protein products coded for by the inserted genes.

One dream, for instance, which has now been fulfilled, was to clone human insulin genes in an easily cultivated microorganism and manufacture the protein, insulin. Even more exotic possibilities were envisioned: synthesizing vaccines against hepatitis, herpes, rubella, and other diseases; making therapeutic proteins, such as antihemophilic factor; making silk (a protein) without using silkworms; repairing or augmenting human chromosomes directly.

At the same time, though, these discoveries induced fears of accidental, naïve, or malevolent production of harmful hybrid organisms; crossing species barriers and tampering with evolution; and in the long run, disturbing or wrongfully manipulating human breeding.

Phases in the Controversy

As with any history, one can see in the recombinant DNA controversy intricate personal dynamics and multiple social causes.[10] I will précis the landmark events, provide a synopsis of the arguments, then comment.

The first phase in the affair was one of question raising and precaution. In 1971, at a routine conference at Cold Spring Harbor Laboratory on Long Island, Janet Mertz of Paul Berg's Stanford laboratory discussed a proposal to clone the much studied tumor-inducing virus SV40 in a laboratory mutant of the bacterium *Escherichia coli*, the benign denizen of the human gut. Some observers, such as Robert Pollack of the Cold Spring Harbor Laboratory, reacted with alarm. Although SV40 itself was thought not to be capable of infecting humans and causing cancer (its tumorigenicity was confirmed only for rodents), was there a chance that the SV40 genome would gain human infectivity by being conveyed in Trojan-horse *E. coli*? After consulting colleagues elsewhere, Berg postponed the experiment. A small group met at the Asilomar Conference Center in California in January 1973 to discuss the hazards of experimenting with animal viruses. Debate intensified.

The next phase brought amplification of concern. Apprehensive about the potential of the newly described restriction enzyme and plasmid tools, in the closing session of a Gordon Research Conference in New Hampshire in June 1973 conferees held an inconclusive discussion, but drafted the dramatic letter to the presidents of the National Academy of Sciences and the Institute of Medicine that was quoted at the opening of this section. *Science* and its British counterpart, *Nature,* reprinted the letter. Later James Watson would comment (Watson and Tooze, 1981, p. 3):

> These 48 molecular biologists took upon themselves the responsibility for announcing publicly, without further private discussions within appropriate professional bodies in the United States and elsewhere, their unease about the direction in which molecular biology was moving. Whether or not a vote on this issue, in which a larger proportion of the world's molecular biologists and medical microbiologists had been enfranchised, would have led to the same result is anybody's guess.

A National Academy of Sciences committee, chaired by Paul Berg, was formed to consider the problem, and in July 1974, via a press conference

10. Good chronologies of the early controversy are U.S. Congress, Congressional Research Service, 1976, and Swazey, Sorenson, and Wong, 1978. Watson and Tooze, 1981, reprints many documents, with idiosyncratic commentary. Good general references are Wade, 1979; Jackson and Stich, 1979; and Morgan and Whelan, 1979. Krimsky, 1982, covers much anecdotal material but is uneven in analysis. U.S. National Institutes of Health, Office of the Director, *Recombinant DNA Research* (multiple volumes), is the primary documentation of the work of the Director's Advisory Committee 1976–1978.

and a letter to the editor of *Science*, it called for a voluntary worldwide moratorium on certain of the more risky-seeming experiments, urged caution in general, called on the National Institutes of Health (NIH) to establish an advisory committee and devise guidelines, and announced another Asilomar meeting—all based on "judgments of potential rather than demonstrated risk" (Berg et al., 1974). All accounts indicate that in the United States the moratorium was observed. In Great Britain the government's Medical Research Council instructed scientists to observe the moratorium called for in the Berg committee letter and chartered a working party (Ashby committee) to develop recommendations. The European Molecular Biology Organization discussed the issue but did not establish any policy.

Next came a phase of argument within the research establishment. Berg and other leading biological scientists organized an "experts' town meeting" at Asilomar in February 1975, with financing by the National Science Foundation and the National Cancer Institute. The meeting brought together 155 invitation-only participants, comprising 83 American research, governmental, industrial, and legal conferees, 51 representatives from Europe, Japan, Australia, Israel, and the Soviet Union, and 21 lay and newsmedia participants. Discussion ranged widely over the technical prospects for recombinant molecule research, risk and benefit guesstimation, legal issues, and freedom of inquiry, and culminated in a hectic all-night session of guideline drafting and debate.

The Asilomar conference is often pointed to, now, as the first occasion—except perhaps for the atomic bomb project—on which a community of scientists voluntarily took initiative to critique consequences of its own research. Attendees have described the meeting as vigorous, disturbing, and chaotic at the end, and they have ascribed to participants motivations ranging from self-serving interest in lifting the moratorium and getting experiments moving again, to genuine fear of hazard, to a wish to defuse public and government concern. Five years later DeWitt Stetten, NIH deputy director for science, would recall Asilomar as having "had many elements of a religious revival meeting. I heard several colleagues declaim against sin, I heard others admit to having sinned, and there was a general feeling that we should all go forth and sin no more" (1978, p. 206). In including the chiefs of most of the laboratories pioneering in this research in the United States, and in including leaders of government R&D and regulatory programs, and in involving the National Academy, the British Medical Research Council, and their counterparts from other countries, the conference was quite establishment in character. Eventually a summary was published in the *Proceedings of the National Academy of Sciences* and in *Science*.[11]

Then came striving for broader consensus. An NIH Recombinant DNA Molecule Program Advisory Committee (nicknamed RAC), which had

---

11. Berg et al., 1975. Rogers, 1977, is a journalist's account of Asilomar and later events.

been formed in the October before Asilomar, drafted guidelines for laboratory practice, based partly on the Asilomar conference's recommendations. Researchers pushed to develop handicapped strains of bacteria that could be employed as laboratory hosts but relied upon not to survive or infect if they escaped containment. Senator Kennedy's Subcommittee on Health held an exploratory hearing on "The relationship of a free society and its scientific community," centered on the recombinant DNA problem. The NIH advisory committee (RAC) was broadened to include members of the lay public "knowledgeable in the fields of ethics and socioeconomics as well as biological research"; these included such extraordinary lay people as Peter Hutt, former general counsel of the Food and Drug Administration and attorney for many bioindustrial firms, and David Bazelon, chief judge of the Federal District Court of Appeals for the District of Columbia circuit.

In July 1976 the final NIH Guidelines were published in the *Federal Register* (41, pp. 27,902–27,943). The Guidelines expressed general admonitions and specified levels of containment for experiments of different estimated levels of hazard. Physical containment ranged from P1 minimal (specifying, for example, "When pipetting by mouth, cotton-plugged pipettes shall be employed"), through P2 low ("Liquid wastes of recombinant DNA materials shall be decontaminated before disposal"), through P3 moderate (laboratories must have "special engineering design features and physical containment features" and "only persons whose entry into the laboratory is required on the basis of program or support needs shall be authorized to enter"), to P4 high ("[experiments must be performed only in facilities] designed to prevent the escape of microorganisms to the environment, [including] monolithic walls, . . . air locks, . . . double-door autoclaves, . . . and a separate ventilation system"). Biological containment ranged from EK1 moderate ("*E. coli K-12* hosts with pSC101 plasmids"), through EK2 high ("host–vector systems that have been genetically constructed [so as not to permit survival] in other than carefully regulated laboratory environments at a frequency greater than 10^{-8}"), to EK3 very high ("EK2 systems confirmed by independent tests in animals . . . and approved by the NIH").

The next events can only be described as the "thrashing-out- the-issues" phase. The NIH Guidelines and associated regulations became the target of much debate, in part because the specific provisions lent themselves to criticism, in part because as the dominant federal health R&D agency the NIH was funding much of the recombinant DNA research, and in part because in recent years the NIH had been the focus of general bioethics discussion.

The National Academy of Sciences held a public forum on recombinant DNA in 1977, but the Academy did not perform major studies on any aspects of the problem, probably because its leaders thought the NIH a more appropriate arena. Congressional bills to regulate the research were fought down by the scientific community as not being appropriate or effective instruments. The congressional Office of Technology Assessment avoided the issue entirely until around 1979. Congress considered imposing strict civil liability, as in Senate Bill S.621 of 1977, which would have provided that "Persons carrying out research involving recombinant DNA shall be strictly liable, without regard to fault, for all injury to persons or property caused by such research"; but the bill was not passed. Acrimonious debate raged in such university towns as Ann Arbor, San Diego, Cambridge, and Princeton over whether local regulations should be instituted to supplement the national guidelines. University administrations got involved. Prominent scientists published their views independently. And the public press carried many articles.

The next phase, from 1977 on, was tendentious but brought maturation. Prospects were sized up. Consensus congealed. Polarization diminished. The recombinant DNA issues became integrated into established public health and R&D policy considerations. Although the problem was not over, it became normalized.

Crucial reassurance was secured from the field of medical microbiology, an experienced discipline consulted surprisingly late. Experts scrutinized the records of past bacteriology laboratory accidents, fed *E. coli K-12* cocktails to volunteers and analyzed their stools for evidence of intestinal colonization, and reviewed evidence as to whether plasmids could shuttle genes into the cells of higher animals. Most experts judged the contemplated recombinant-gene experiments unlikely to produce organisms capable of effecting the sequence: surviving outside of laboratory culture, infecting higher organisms, colonizing these hosts, and then causing disease either by modifying the hosts' genetic makeup, altering the constitution or activity of other microorganisms inside the hosts, or secreting harmful materials.[12] Molecular geneticists quickly adopted these reassurances.

After that the issue moved on through NIH Guideline revisions, punishment of several minor infractions of the Guidelines, international discussions, more congressional inquiries, patentability suits before the Supreme Court, and the cloning of biotechnology firms.

The Issues Recapped

Looking back over events, it is obvious that the controversy had to do with "responsibility," but that inherently it wasn't clear who was obliged to be responsible to whom for what. The controversy had to do with freedom of inquiry, but few researchers asked for absolute freedom, and they acknowledged that many other factors were at stake. The controversy had

---

12. Gorbach, 1978, was an especially influential conference.

to do with public health threats, but these were so speculative that physicians and other traditional health professionals played little role. This is why I think it best to recognize the whole affair as a renegotiation of the social compact surrounding this area of science. The early phase, roughly 1971–1978, is the one that concerns us here. I will paraphrase what I believe were the central questions and consensus answers.

The first issue to surface was that of risks. Are the methods themselves dangerous to perform? (The answer, coming quickly, was, Not apart from the particular products the procedures make.) Could some imaginable recombinant products be harmful? (The answer became, Yes, possibly: for example, a normally benign bacterium carrying transplanted genes for bubonic plague toxin could, for all we know, cause disease.) How harmful are they likely to be? (We have no way of knowing for sure. Key experiments have to be performed first. Experts in communicable disease will have to be consulted.) Are the risk-probing experiments themselves hazardous? (Hard to say; they should be performed carefully.) Is there risk from laboratory accidents? (Possibly; but probably no worse than from other experiments with viruses, toxins, and radiation, that go on, under control, and to great benefit, all the time.) Can the risks be categorized as to degree of threat? (Yes, but not precisely. This needs to be discussed widely among different scientific communities.) Couldn't the techniques be abused by people of ill will? (Sure. But given the myriad explosive, poisonous, biological, and other disruptive possibilities already available, these arcane techniques would hardly enlarge the arsenal.) Won't these techniques lead to manipulation of Nature? (Some will. But spontaneous mutations and plasmid-mediated gene exchanges occur naturally all the time. We manipulate Nature constantly. Besides, much of the recombinant DNA research is intended to provide remedies for handicaps, diseases, and blights levied on us *by* Nature.)

Risk discussions tended to be naïve. Then, as even now, it was impossible to estimate risks precisely, just as, to mention two cases not related to recombinant DNA, it is impossible to anticipate how virulent slightly modified strains of influenza will turn out to be (Hong Kong flu, swine flu . . .), or to understand how it was that very small mutations in *Staphylococcus aureus* gave rise to the strain that causes toxic shock syndrome. It is regrettable that some recombinant DNA veterans now pooh-pooh the Asilomar apprehensions. The public had every reason to be concerned. So did the experts: there was nothing—nothing—in empirical science in the early 1970s to allow reliable estimation of potential harm from these experiments. Molecular geneticists were not themselves able to assess the transmissibility and pathogenicity of even *E. coli*.

My own biological intuition is that most recombinations are extremely unlikely to create organisms capable of harm. However, although it is reassuring to be able to relax over *E. coli* experiments, countless other manipulations are possible. As we have learned from the way the commercially hybridized "killer bees" that escaped in Latin America have so dangerously spread and stung their way northward, even mundane breeding can lead to harm. As we have learned from DDT-resistant mosquitoes and penicillin-resistant gonococci, changes in biotic selection pressures can have serious consequences. Even if American scientists obey guidelines, others in the world will not necessarily be constrained. And industrial-scale cultivation of recombinant strains may bring hazards not encountered at laboratory scale. Some risk will always exist.

As counterbalance to the concerns about risks, there have always been visions of benefits: rich new basic knowledge; production of novel and scarce hormones and antibiotics; direct hybridization of plant species; eventually, splicing missing genes into deficient human chromosomes; and—who knows? The potential benefits were obvious enough to induce a rash of incorporations, runs on the Patent Office, and splashy stock offerings.[13]

In succeeding chapters I will discuss the generic issues of freedom of inquiry and of public participation, both of which pervaded this debate but are not unique to it.

Process Problems

In the recombinant DNA negotiations what most scientists and their institutions were asking society for, and got—painfully and tentatively—was permission to proceed, assurance that they could have self-determinative leeway, and continued commitment of resources. No new laws were passed or strict regulations imposed. In return, scientists promised benefits, self-control within guidelines set by institutions such as the NIH, and acceptance of the understanding that breaches of the compact, whether accidental or deliberate, likely will trigger strict regulatory sanctions.

Thus responsibility was assumed voluntarily and communally, but under scrutiny by the public and government. The alarums, self-moratorium, Asilomar debate, guideline drafting, and other initiatives may have been self-serving, but in many ways they manifested exemplary stewardship. The affair provoked several ill-defined groups into reorganizing themselves under the new "recombinant DNA R&D" rubric, developing communal stances on the issue, and cultivating relations with the established communities of communicable disease research and fermentation engineering.

Traditional sources of moral guidance never were tapped deeply. A few religious groups passed resolutions calling for moratoria on research. The same several lawyer-observers who usually get involved in such matters did so, but most law review articles on the issue dealt only with constitutional protection of freedom of research, or with tort liability. Ethicists wrote

13. U.S. Congress, Office of Technology Assessment, April 1981, surveys prospects.

articles pointing out that indeed new ethical questions were involved, but they offered hardly a shred of useful guidance, except on social procedure.[14]

It is very instructive now to think back over the early years of the recombinant DNA controversy, canvass what key participants think about the dynamics, and ask how the events could have been handled better. One review, for instance, by Sheldon Krimsky, concluded that the Asilomar conference could have been improved (1982, p. 151):

> An alternative way to convene a conference of this nature might have included: contacting the professional associations that had the expertise in a relevant area (infectious disease, medical microbiology, immunology) for nominations of participants; an open request for papers on the potential risks of using certain hosts in the cloning experiments; contacting environmental organizations for their expertise in problems of monitoring agents that are disseminated in the environment; soliciting participation from agencies and organizations concerned about occupational health. Once these inputs were available, working panels could have been established with a wider range of representation from the scientific community and public interest groups.

These kinds of recommendations deserve wide discussion. Inevitably the future will bring other episodes of this kind.

Most veterans of the controversy express great weariness. Their reflections are divided. Looking back in 1980 Norton Zinder of The Rockefeller University despaired, "I doubt that any group of scientists would now speak out on an issue with only potential hazard" (1980, p. 15). But on that same occasion MIT's David Baltimore said, "There are very few groups in this society who are willing to question themselves. We did it once on recombinant DNA and I hope that we are willing to do it again" (1980, p. 15).

---

14. A substantial symposium, albeit one with a restrictionist cast, was *Southern California Law Review*, 1978.

Six

Societal Guidance
of Inquiry and Application

Far more than many commentators acknowledge, society—the public and its leaders—exerts diverse and powerful compactual leverage on technical activity. Directly and indirectly, society brings its will to bear on: the general complexion of R&D activity (as with agricultural research priorities, or whether the Landsat program will remain a civilian program or become military); professional credentialing and practice (licensing of engineers, or appointment of food-and-drug regulators); quality of technical assessment and advice on decisions that affect the public (use of scientific judgments by the Environmental Protection Agency); criteria for professional judgments made in service of the state (definition of death); conditions under which new knowledge is gathered and used or experiments conducted (diplomatic protocol for handling archaeological finds, or protection of prisoners in psychiatric experimentation); undertaking of new lines of technological development (whether to develop satellite weapons); possession and transfer of knowledge or know-how (patent protection of new life-forms, or proposed Law of the Sea mandate on sharing of seabed mining technologies); and conditions for deployment of technologies (where to build liquefied-natural-gas terminals, or proper use of genetic screening).

These controls, along with the increase in politicization I have been describing all along, have made major incursions into the freedom of technical activity. Earlier chapters have described such incursions as restriction of oceanographic and archaeological access, lawsuits against voluntary engineering standard-setting, restraints on release of public health data, and increasing regulation of medical and R&D practice in general. As in all of life, freedoms and regulations exist in mutual tension.

The first part of this chapter will explore the notion of scientific freedom. The second part will examine some of the mechanisms and controls through which society guides technical activities.

HOW FREE SCIENTIFIC FREEDOM?

Although disputants often appeal to the ideal of "scientific freedom" during quarrels such as that over recombinant DNA research, both the philosophic foundations and the specific implications of that freedom are vague.[1]

The proposition that scientists should be free to pursue their inquisitiveness without undue interference by the laity rests on three grounds. The first is the tradition of scholarly freedom spawned by the Renaissance struggle to desacralize aspects of order, causality, and teleology, and allow science to form truths independent of religious tenets. But traditional freedoms are only traditions, and this one has proven easily suspendable during such nasty episodes as the Weimar *Rassenhygiene* movement, Lysenkoist dictatorship of Soviet agricultural genetics, Nazi perversions of eugenics and medicine, and a variety of more recent totalitarian shames (Loren Graham, 1981). "Academic freedom" obviously is a related notion. But universities now are run by state legislatures and other public trustees; taxpayers pay much of the research bill; and, besides, a lot of potentially contentious technical activity is conducted in industrial, government, and other laboratories outside of academia, often under the privity of corporate or military security. It is asking too much to expect traditional academic freedom, which relates mostly to the educational function of universities, to extend over all this diverse research activity.[2]

In the United States, a second ground is constitutional protection under the First Amendment. But while freedom of speech has been interpreted by the Supreme Court as guaranteeing freedoms of educational expression, freedom of research inquiry has not been defended so stoutly.[3] As with all First Amendment issues, conflicting considerations often arise. Social scientists may, for example, find themselves treated like journalists and others whose First Amendment rights to gather information have been held not to outweigh other considerations, such as protection of "private

1. Essays on scientific freedom include American Association for the Advancement of Science, 1965 and 1975; Holton and Morison, 1979, which is a reprint of *Daedalus* 107, no. 2, 1978; Delgado and Millen, 1978; Robertson, 1978; and Wulff, 1979.

2. Metzger, 1978, discusses scientific freedom as a dubious extension of academic freedom.

3. *Sweezy* v. *New Hampshire*, 354 U.S. 234, 250 (1957), addressed state legislative review of university lecture content. *Keyishian* v. *Board of Regents* [*of New York*], 385 U.S. 589 (1967), reviewed political loyalty oaths for teachers. *Epperson* v. *Arkansas*, 393 U.S. 97 (1968), examined the teaching of Darwinian evolution, and *McLean* v. *Arkansas Board of Education*, 529 F. Supp. 1255 (E.D. Ark. 1982), the equal-time teaching of biblical "creation science."

space," confidentiality of jury deliberations, or the federal government's authority to learn informants' names, sequester prisoners, deny passports, or protect survey subjects from mental harm.[4] The principles of the American Constitution are not likely to be defended analogously elsewhere—which in this day of multinational corporations and international R&D institutions seriously weakens the case.

The third ground, and the one I consider most defensible, is the autonomy established by the compacts various fields have negotiated between themselves and society. What is implied is the bargain reflected in the title of the principal report by the American Association for the Advancement of Science on the issue: *Scientific Freedom and Responsibility* (1975). Like other creative people, scientists tend to make progress best when they are unfettered. Probably the strongest reason for championing the hallowed ideal of scientific freedom is, quite pragmatically, that it gets scientific results. But for many obvious reasons the R&D process and results must be subjected to some social scrutiny and control. (As scientists might express it in analogy to electrical circuitry, there must be an "impedance match" between technical progress and social progress.) Within a compact, bounded independence can be negotiated that can be specific, tailored to the peculiarities of the field, and made durable by societal oversight guarantees.

Often scientific freedom is construed as extending through a supranational fraternity of scientists. This high principle is cited in efforts to protect scientists against totalitarian repression, and in condemning psychiatrists and other technical experts who abet that repression. Worthy though the end may be, however, appeal to such insubstantial principle is not very compelling. The following statement, made on behalf of the American Association for the Advancement of Science's Committee on Scientific Freedom and Responsibility, is about as firm a rationale as can be presented (Eisner, 1983):

> We work on behalf of scientists because we share a professional identity with them and because we feel that they are often targeted for repression due to their professional visibility. This is not to say that

---

4. *Le Mistral, Inc.* v. *CBS*, 61 App. Div. 2d 491, 402 N.Y.S. 2d 815 (1978), held that a television station had violated a restaurant's privacy by entering without permission and filming health code violations. *Branzburg* v. *Hayes*, 408 U.S. 655 (1972), tested journalists' privilege to maintain confidentiality of informants. *Pell* v. *Procunier*, 417 U.S. 817 (1974), and *Saxbe* v. *Washington Post Co.*, 417 U.S. 843 (1974), upheld state and federal prohibitions of media access to prison inmates, even those willing to be interviewed. *Zemel* v. *Rusk*, 381 U.S. 1 (1965), upheld the State Department's denial of passport protection for Zemel to travel to Cuba to gather firsthand information. *Trachtman* v. *Anker*, 563 F.2d 512 (2d Cir. 1977), *cert. denied*, 435 U.S. 925 (1978), ruled that a school had a right to forbid a highschool journalist to administer a voluntary questionnaire to fellow students about their sexual attitudes and practices, on grounds that the school board had demonstrated potential for emotional traumatization of some respondents. Robertson, 1978, traces the path through most of these cases.

they are more sensitive to political oppression or more conscionable and outspoken, and for that reason likely to be targeted for harassment or persecution. Scientists share their discoveries and views with a community that is international. They pose a threat to totalitarian regimes because their criticisms may be heard worldwide.

The issue, which lies beyond the scope of this book, is important and needs critical study.[5]

As to whether limitations ever should be placed on research, extreme opinions occasionally have been voiced like this one from National Academy of Sciences president Philip Handler, on cloning (1969):

> [NAS] NEWS REPORT: Do you feel any constraints should be placed on fundamental research if there is some reason to believe that the results of that research might be harmful to society?
>
> HANDLER: No. No constraints. Let me give you a dramatic illustration. You may know of the demonstration that one can take a fertilized frog egg, discard its nucleus and insert a nucleus from a somatic cell of some other frog, and the egg develops into a frog which is an absolutely perfect twin of the donor frog—the one that provided the transplanted nucleus. Presumably, by that technique we would make an indefinite number of perfect copies of that donor frog. It's merely a matter of time before we can switch from frogs to mammals. Obviously, the next step would be man. . . . I hope that day never comes. I can't imagine any more dangerous tool in the hands of an autocratic, dictatorial, authoritarian government. . . . And yet I think there is no alternative but to go down this trail and do the biological experimentation that, one day, may offer this kind of capability. . . . No constraints.

The rhetoric of some science spokespersons knows no bounds. In a moment of overreaction during the recombinant DNA brouhaha in 1976 Gerard Piel, the publisher of *Scientific American*, issued the pronunciamento, as addled on scientific autarchy as on citizenship: "A scientist can accept no authority but his own judgment and conscience, just as a sovereign citizen must be an autonomous self-governing member of society" (1976, p. 19).

Unconstructive nonsense. Nobody, not even (the late) Handler or Piel, believes that scientists should be allowed to do just any experiment they want to, absolutely without regard for consequences or costs. *Pace* Bacon, nobody really wants the "effecting of *all* things possible." (As George Bernard Shaw nailed it in the preface to *The Doctor's Dilemma*, "No man is

5. One review, based on international human rights law, was Council for Science and Society [U.K.] and the Governors of the British Institute for Human Rights, 1977.

allowed to put his mother into the stove because he desires to know how long an adult woman will survive at a temperature of 500° Fahrenheit, no matter how important or interesting that particular addition to the store of human knowledge may be.")

The label "pure science" can be overly sanitizing. As happened with oral contraceptives and lasers, fundamental scientific discovery can rapidly lead to technological exploitation. Moreover, in these cases as in most others, the general potential for application of the endocrine research and the electro-optical experiments was quite evident to the researchers and their patrons long before the research succeeded or the ramifications became clear in detail. Programs of basic science, if not individual researchers' projects, almost always are pursued with some—even if distant—humanitarian, commercial, or military goal in mind.

The difficulty in controlling research is that neither scientists nor anybody else can know the outcome of experiments in advance, or imagine how diverse research themes might suddenly intersect, or anticipate all possible physical and social consequences of change. The very nature of discovery and development, in which fairly unpredictable spurts of advance alternate with incremental growth, limits the precision with which constraints can be applied. Research is conducted in so many places that restraining it in one locale almost surely will not stop it everywhere, although cautious example may influence the character of development at other centers: even total prohibition of recombinant DNA research in the United States probably would not have prevented its being pursued, in the long run, elsewhere. Even less valid than assertions of unrestrained *scientific* freedom are such claims of *technological* freedom as Oppenheimer's absolution of the atomic bomb project: "When you see something that is technically sweet, you go ahead and do it and you argue about what to do about it only after you have had your technical success." This was fatuous even when he said it to the Gray Board in 1954 (U.S. Atomic Energy Commission, 1954, p. 81).

Should any research ever be forbidden in advance because of the knowledge it potentially might generate? Only very rarely, we should hope, excepting atrocities that already are demarcated by morality and law. But we must not close off the possibility.

One case of opposition on prior principle arose in the 1970s regarding research on whether being born with (rare) XYY chromosomes predisposes men to violent or criminal behavior. Critics charged that such research could lead to prejudicial screening programs: not every XYY male would necessarily be antisocial, and informing boys and their parents of their extra chromosome would invade privacy, stigmatize the boys, and tend to induce self-fulfilling sociopathic behavior. Proponents of screening argued that the knowledge would be advantageous in providing early warning. Debate raged for several years, but then faded as the behavioral assessments were judged to be too multifactorial and uncontrolled to be conclusive, thereby mooting the issue. Even in this example, though, apprehensiveness concerned not just "pure knowledge" but its potential application to social screening (Gaylin, 1980).

Almost always the question is not whether to avoid a whole line of exploration, but when, and under what experimental conditions, and following what preliminary assessments, and under what assurances of social trust and accountability, research should reasonably proceed. *Reckless*—"reckless"—science, as Jerome Ravetz has termed it, surely is what we want to avoid (1971, p. 56). Recall the recombinant DNA case. Prohibition was proposed early along for some experiments, both out of apprehension about particular hybridizations and out of concern that the cumulative agenda would amount to "playing God." As it worked out, some experiments were forbidden under the Asilomar and NIH guidelines, others were held back until pilot assessments of infectivity and other surrounding issues could be conducted, a few controversial ones were conducted under rigorous precaution, and a variety of social assurances were exchanged (between local communities and universities, for example) and controls instituted.

As to pacing of controversial research, Robert Sinsheimer has provided the apt term by suggesting that on occasion we may want to forbear "inopportune knowledge," knowledge "the possession of which, at a given time and stage of human development, would be inimical to human welfare" (1978, p. 23). Loren Graham has nominated this provocative case (1978, p. 21, footnote 18):

> Imagine that you are a scientist in Nazi Germany and that you have just discovered Tay-Sachs disease, an abnormality based on a genetic defect which is more common among Jews than other population groups. Is it not possible that you would try to keep this bit of knowledge away from the eyes of Hitler and National Socialist bureaucrats, and that you might suggest to your trusted colleagues that they do the same?

Impending imaginable real examples are human–animal chimeras and human pheromones. By manipulating embryonic cells from different species, biologists recently have generated goat–sheep chimeras that grew to adulthood; the prospect of hybridizing humans with nonhuman species is distant but not impossibly farfetched (Dixon, 1984). Some mammals secrete pheromones (hormone-like chemicals transmitted in minute amounts via air or water which deeply influence the behavior of other organisms) that inhibit puberty, block pregnancy, cause sexual arousal, induce fear and cowering, and exert other powerful effects. There is some evidence of pheromones in humans. Wouldn't we want to proceed very cautiously in isolating and synthesizing human pheromones? The potential for subliminal social control, or for sexual abuse, could be enormous.[6] Surely, someday soon, some such biomedical or other research proposal will be judged as holding enough antihumanitarian potential to warrant restriction.

SOCIETAL GUIDANCE OF TECHNICAL ACTIVITY

Political and Financial Support of R&D

Although we may prefer never to ban research, in no way are we able or obliged to support all lines of research equally. And even though the practical potential of fundamental research never can be predicted fully, much can be foreseen, especially the likely consequences of forgoing particular projects: if we don't commit resources to studying controlled nuclear fusion, we will never understand fusion or be able to use it for making electricity. Funding-choices obviously influence technical advance.

The relation of fundamental or basic research to application is complex and not entirely predictable. An example will illustrate. In a classic study, Julius Comroe and Robert Dripps analyzed key publications about major clinical advances made during the period 1945–1975 in cardiovascular and pulmonary medicine. One conclusion was that dramatic landmark advances, such as open-heart surgery, had drawn heavily on many previous, disparate discoveries: open-heart surgery could not have been achieved without electrocardiography, blood typing and preservation, anesthesia, surgical techniques, wound and infection control methods, and many other advances—each of which in turn had drawn on many previous, disparate discoveries. Another conclusion was: "Of 529 key articles, 41 percent of all work judged to be essential for later clinical advance was not clinically oriented at the time it was done; the scientists responsible for these key articles sought knowledge for the sake of knowledge."[7] From this and similar reviews of technical innovation we understand the importance of pursuing basic research. Also we realize that such research, though fundamental, must not necessarily be "blue sky" knowledge-for-its-own-sake research but may be oriented to identified needs. (And we infer, inversely, that trying to exert social control over any development having as many different origins as open-heart surgery will likely be very difficult.)

From time to time scientists feel it necessary to remind the world of the unpredictability of research, of the difficulties for public understanding of arcane work, of the educational and other spin-off benefits that accrue from academic scientific activity, and of the desirability of maintaining

6. Vandenbergh, 1983, reviews recent pheromone research.

7. Comroe and Dripps, 1976. Two other studies that trace lines of innovation are Illinois Institute of Technology Research Institute, 1968, and Batelle Columbus Laboratories, 1973.

national technical competence. But then having done so they may go further, as Harvard physicist Steven Weinberg did in 1974, and assert that "in seeking support for scientific research, scientists need not agree with the public as to why the work should be done" (1974, p. 37). An extreme version of this argument was made by Jacob Bronowski, who called for the "disestablishment of science, in the same sense in which the churches have been disestablished and have become independent of the state," and said that "the time has come [1971] to consider how we might bring about a separation, as complete as possible, between science and government in all countries," because "the choice of priorities in research should not be left in the hands of governments" (1971). No one heeded Bronowski's call, which surely was a futile one.[8] Weinberg's slightly overdrawn complaint deserves to be recognized; it must be answered by recommendations of mechanisms through which basic science can be supported from the public purse on faith as a generally-good-thing, but through which the broad priorities and conditions of research are subjected to public guidance.

Popular concern over priorities is encapsulated in such laments as, "We are the generation that put a man on the moon while standing knee-deep in garbage"; and "Great, the supersonic airliner will enable us to go from Watts to Harlem in three hours"; and "If this country can build nuclear submarines, can't it make quieter garbage trucks?"; and "Why no male contraceptive pill?" The difficulty, of course, is that R&D options are almost never intercompared so directly. And some things, such as intervening in male fertility, we simply haven't figured out how to go about doing.

To analyze and influence government and corporate R&D priority-setting tends to be very complicated. Resource allocators usually take into account both client demand (sales-market demand, or electoral constituency wishes, or internal needs within the organization) and scientific community pressure both for knowledge-for-its-own-sake and for potentially applicable knowledge. Broad-scope priorities can be discerned—the United States either supports the Voyager mission to Saturn or it doesn't—but smaller, more abstract, or less definable ones may be much harder to factor out. To analyze the pattern of all federal research on, say, health effects of asbestos would be a major project, and that would not include the substantial research being done outside of government auspices.[9]

Since 1976 the American Association for the Advancement of Sciences has rendered major service by preparing analyses of upcoming federal R&D budgets, which it rightly describes as being "derivative of a highly

8. Disapproving reactions by Peter Medawar, Gerard Piel, Anthony Benn, and Eugene Rabinowitch were carried in *Encounter* 37, no. 3, pp. 91–95, 1971.

9. Regarding federal R&D priorities as a reflection of public and science community preference, see Brooks, 1978.

complex and somewhat arcane decision-making and bargaining process embedded in a maze of institutions involving both the Presidential and the Legislative systems." This project has provided an important forum, and it has helped educate scientists about the budgetary process.[10]

When the public wants something of science, it tends to say so. Under public pressure, officials at many levels of governance are calling for solutions to the acid rain blight. Concern over arthritis, which is intensifying as the American population ages, has led to lobbying for increased funding of federal arthritis research and for establishment of a new arthritis institute within the National Institutes of Health. The market demand for cheaper personal computers has induced a frantic R&D race, as has the public plea and commercial potential for a herpes vaccine and cure. In response to public concern we have reduced our manufacture and release of persistent chemical pesticides, chlorofluorocarbon spray propellants, narcotics, toxic heavy metals, and such hazardous manufacturing components as beryllium, polychlorinated biphenyls, and asbestos. Steadily we have reduced the average person's x-ray exposure. We are rejuvenating the options of home birth, natural birthing, and midwife delivery, and have been involving fathers more in the process of pregnancy and childbirth. We are applying technology to the preservation of artworks and archival treasures, and to the structural rescue of Venice, the Parthenon, and the Statue of Liberty.

Over the years many major technical projects have been opposed and stopped or modified. Scientific and public pressure helped bring an end to atmospheric testing of nuclear weapons. Political and scientific community resistance stifled Project Orion, the attempt to design a spacecraft propelled by nuclear explosions.[11] The Soviets have refrained from deflecting Siberian rivers into the sea, as has the United States from digging a sea-level Panama Canal. The United States has held off on using nuclear explosives in civil engineering projects even though the Soviets have not, has opposed building the supersonic transport aircraft, and has moved slowly on recycling nuclear fuel and on building fast-breeder reactors. Any of these may, someday, under some conditions, be pursued. But they were not inevitably pursued as proposed.

Regulation of Medical Experimentation

In his 1865 *Introduction to the Study of Experimental Medicine* Claude Bernard proclaimed: "The principle of medical and surgical morality consists in never performing on man an experiment which is harmful to him to any extent, even though the result might be highly advantageous to science,

---

10. These analyses are reported in the AAAS's annual publication, *Research and Development in the Federal Budget*.

11. Dyson, 1965, and Lambright, 1967, describe the project's history 1958–1965.

i.e., to the health of others" ([1865] 1950, p. 101). But surely even then, as now, things were nowhere near so clear-cut, because inevitably some people will be harmed by some tests. Severe risks are entailed even in the practice of conventional medicine. Whether any particular experiment should be conducted must be determined by whether expectations of benefit to the subjects or society outweigh the risks to the subjects or society, and by whether other ethical considerations are satisfied.

The era of medical-subject-as-martyr came to an end after World War II, stimulated in part by the Nürnberg tribunal's revelations of Nazi torture and pseudoexperimentation (U.S. Army, Adjutant General's Department, 1947). Leading medical bodies adopted codes of practice. The reforms have been substantial. Now most codes of ethical experimentation require that:

- Preliminary assays on nonhuman species must show promise of human benefit sufficient to justify the experiment;
- Experimenters must fully disclose the nature of the test to the subjects;
- Prior voluntary, informed consent must be secured from the subjects;
- Investigations must be closely supervised by certified physicians and other professionals;
- Every reasonable precaution against mishap must be taken; and
- Both the subjects and the investigators must remain free to stop the experiment at any time.

In recent years a variety of government and institutional oversight mechanisms have been established. In 1973 Senator Edward Kennedy's Senate Health Subcommittee reviewed aspects of pharmaceutical testing and prescription, abuses such as the Tuskegee Syphilis Study on uninformed prisoners, and some failures of physician and institutional review to protect subjects adequately (U.S. Senate, Subcommittee on Health, 1973). From 1974 to 1979 the National Commission for the Protection of Human Subjects of Biomedical and Behavioral Research developed very helpful guidance on such issues as experimentation on living fetuses, aborted fetuses, prisoners, children, and the mentally infirm. Many of the Commission's recommendations have since been promulgated as regulations by the Department of Health and Human Services.[12] After 1979 the President's Commission for the Study of Ethical Problems in Medicine and Biomedical and Behavioral Research continued the work of examining human research regulation.[13]

12. U.S. National Commission for the Protection of Human Subjects of Biomedical and Behavioral Research, various dates. The regulations appeared as U.S. Department of Health and Human Services, 1981. Hutt, 1978, pp. 1455–1469 is an approving review of the Commission mechanism.

13. U.S. President's Commission for the Study of Ethical Problems in Medicine and Biomedical and Behavioral Research, *Implementing Human Research Regulations*, March 1983.

Medical experimentation now is guided by codes of practice, scrutinized by granting agencies, overseen by institutional review boards comprising both technical and public members, and influenced by opinions of professional organizations, bioethicists, and public leaders. The most important changes have been the heightening of sensibilities among researchers and their institutions, the according of much greater respect to subjects' dignity and self-determination, and the development of formal public accountability.

Many ethical difficulties remain: regarding what constitutes *free* and *fully informed* consent; regarding how to distinguish "experiment" from standard medical testing and therapy; regarding the extension of medical ethical principles to cover testing of cosmetics, toiletries, and other consumer products; regarding how to handle control subjects; and regarding experimental use of cadavers, autopsied tissues, and embryos. Some procedural difficulties remain: regarding the protective effectiveness and streamlining of institutional review; regarding military testing; and regarding research conducted outside of government auspices.

Guidance of Social-Science Experimentation

On occasion social-science researchers have done things that in any other circumstances would be considered indecent practical jokes. In stress studies soldiers have been deceived into believing that their airplane was about to crash, or that during training they had caused a comrade's death. In Stanley Milgram's obedience studies participants were induced, at the investigator's command, to administer faked but supposed painful electric shocks to punish inept "learners" whose cries they could hear through a partition.[14] In Philip Zimbardo's jailor–prisoner roleplay experiment student volunteers were placed in a mock prison situation in which "guards" genuinely came to behave sadistically and "inmates" responded with hostility (Zimbardo, 1973). In Laud Humphreys's "tearoom" studies he served as a watchqueen to observe male homosexuals rendezvousing in public restrooms, and later he interviewed them, pretending to be conducting a public health survey, after tracing their home addresses via their automobile license numbers.[15] In altruism studies subjects unknowingly have been confronted with challenges to their willingness to help people out of sham emergencies, and in temptation tests they have been presented with easy opportunities to cheat or steal. Jury deliberations have been bugged. Prostitutes have been deployed as informants. Families' household garbage has been analyzed. Subjects have been implanted with hypnotic suggestions without their consent.[16]

---

14. These experiments are critiqued in Schuler, 1982, pp. 60ff.

15. Humphreys, 1975, reprints the original study and subsequent ethical commentaries.

16. General discussions include Rivlin and Timpane, 1975; Reynolds, 1980; Schuler, 1982; and Beauchamp et al., 1982.

Much relied on have been such deceptive and surreptitious practices as misrepresenting the purposes and identity of the investigators, making false promises, violating subjects' anonymity, employing pseudosubjects, making false diagnoses, using placebos, secretly administering medications and drugs, disguising mental conditioning, and provoking and secretly recording negative behavior (Schuler, 1982, p. 79). Even if employed only passively these techniques can intrude into privacy and erode personal dignity. Postexperiment debriefing, a kind of casual psychological counseling, usually is depended upon to relieve induced stress.

Ethical precedent has been sought from medical ethics, but the social sciences differ from medicine both in the problems they confront and in the protective options available. Prior voluntary informed consent often can be secured, but many behavioral studies would be nullified if subjects were tipped off beforehand as to the exact nature of the investigation. Incompetent or indisposed subjects usually are accorded special care, but, as with mentally infirm subjects or children, proxy or guardian consent must be relied on. Subject anonymity and confidentiality usually are promised, but these sometimes have been violated.

Reforms are being undertaken to minimize deceit, coercion, and duress; to avoid experiments that lack high research promise; to scrutinize experimental practice more thoroughly in general; and to protect personal data much more carefully. In 1981 the Department of Health and Human Services exempted from regulation many types of social-science research (such as those based on surveys, observation of people in public places, and the study of documents), and it required informed consent and institutional review board scrutiny of research having potential to be more intrusive. These regulations govern only projects sponsored by Health and Human Services (U.S. Department of Health and Human Services, 1981).

Other types of social research also raise ethical issues. Questions of consent, the rights of subjects, and the rights of control groups deprived of the special treatment may be raised in social policy experiments on child day-care, housing redevelopment, medical insurance, police patrolling, and family financial income guarantees.[17] In the past some political-science investigations, such as the ill-conceived Project Camelot in the mid-1960s on "the political, economic, and social preconditions of instability and potential Communist usurpation of power" in several Latin American countries, have been judged inappropriate and have been terminated.[18]

Guidance of Other Aspects of Biomedical Science

In an episode that must sound familiar to laboratory administrators even today, in 1884 citizens living near Louis Pasteur's new facility at Ville-

17. Rivlin and Timpane, 1975, provides an overview. Pechman and Timpane, 1975, describes a leading case, the New Jersey negative-income-tax experiment.

18. Horowitz, 1967, and U.S. Congress, Congressional Research Service, 1979, pp. 145–179, chronicle the Camelot debacle.

neuve-l'Étang on the outskirts of Paris protested that the dog colony, used for testing rabies vaccine, would be noisy, present a serious public health nuisance, and "arrest the development of a very promising neighborhood." Pasteur reassured the protesters that he had taken due precaution, and that the research benefits could be expected to outweigh any risks (Académie des Sciences, 1884). This tension has surrounded animal experiments ever since. Similar concerns have been raised about labs doing experiments involving pathogens and radioactive agents. In recent years laboratory animal husbandry and the security of laboratories have been much improved. Concern for the welfare of animals in experiments has led to a variety of reforms, principally minimization of the number of animals tested (as by substituting other methods of assay) and minimization of pain and duress to the animals.[19]

For better or not, many laboratory activities are being regulated. Clinical laboratory practice is regulated under the Clinical Laboratory Improvement Act and the Medicare Act, and accreditation is administered by the Joint Commission on Accreditation of Hospitals, the American Osteopathic Association, and the College of American Pathologists. Pharmaceutical, medical-device, and cosmetics R&D are closely regulated.[20] Instances of fraudulent drug testing in the 1970s led the Food and Drug Administration to establish Good Laboratory Practice Regulations (U.S. Food and Drug Administration, 1978). Currently there is concern about how to institute safeguards against incompetent and fraudulent toxicological testing of pesticides, industrial chemicals, and other environmental agents. There also is concern about fraud in biomedical research in general.[21] The recombinant DNA guidelines are of course an instance of societal control over experimentation, and several violations have been punished.

Control of U.S. National Security Technologies

As in other countries, overt military research in the United States is held under strict national security classification, to prohibit unauthorized access. Hardware and information releases are regulated under several export control acts, the Invention Secrecy Act, and nuclear security acts. Visiting scientists' visas are restricted when judged necessary. Safeguards, such as those promulgated by the Nuclear Regulatory Commission and the International Atomic Energy Agency, are applied to fissionable materials.

19. U.S. House of Representatives, Subcommittee on Science, Research and Technology, of the Committee on Science and Technology, 1981b; and Rowan, 1984.

20. Merrill and Hutt, 1980, is a thorough legal overview.

21. See U.S. House of Representatives, Subcommittee on Investigations and Oversight, of the Committee on Science and Technology, 1981.

In principle most of these controls are as old as war itself. But recent years have brought a restrictionist drift, partly because of concern over dual-use technologies, such as those in microelectronics, lasers, and navigation systems, that may be developed in civilian contexts but that also can be militarily useful. Too, as disembodied know-how is recognized as being strategically important, fairly abstract research, teaching, and information exchange are being viewed as sensitive.[22]

There have been serious debates about whether to permit students from some countries to study in American technical universities, and about how to restrict scholarly research exchanges. In the early 1980s, for example:[23]

> The Department of Energy issued an order requiring government clearance of any communication between DoE contractors and Soviet scientists. The Commerce Department forced the American Vacuum Society to withdraw its invitation to Soviet bloc scientists to attend a conference on magnetic bubble memory devices, asserting that oral exchanges of information with foreign nationals fall under Export Administration Regulations and that sponsors would have to obtain an export license before admitting Communist bloc scientists. Then, the State Department refused to issue visas to eight Soviet scientists who had planned to attend a conference on laser and electro-optical systems.

The National Security Agency has harassed academic cryptology researchers to try to get them to submit their papers to the Agency for prepublication screening; the mathematicians involved have resisted such censorship, although some have voluntarily consulted with intelligence officials before publishing.[24]

For counterexample I would point out that academic scientists who in other situations defend freedom of inquiry and publication have themselves on occasion urged restricting know-how, as when in the late 1970s many urged that research on laser-driven enrichment of uranium, a method probably easily transferred once its practicalities have been worked out, be placed under national security classification so as not to contribute to nuclear weapons proliferation (Casper, 1977).

Obviously these issues are sensitive matters of balance. I agree that we need to maintain a variety of firm controls on information, capabilities, and hardware relevant to international security. But the sordid exerting of

---

22. These matters are reviewed apprehensively in U.S. Department of Defense, Defense Science Board, 1976, and in NAS, Panel on Scientific Communication and National Security, 1982.

23. NAS, Panel on Scientific Communication and National Security, 1982, p. 27.

24. *Science* has covered the issue since 1980. Unger, 1982, summarizes the issue and related matters up to early 1982.

pressure by trenchcoated agents on academic researchers and teachers, the promulgating of bloated lists of "gray" technologies for restriction, and the censoring of scientific publications, in the long run are ineffectual. The National Academy's conclusion is also mine: "The best way to ensure long-term national security lies in a strategy of security by accomplishment, and an essential ingredient of technological accomplishment is open and free scientific communication."[25]

All these societal controls are heavily informed by systematic assessment, to which we now turn, and by the social value judgments to which we will proceed in Chapter Eight.

25. NAS, Panel on Scientific Communication and National Security, 1982, p. 47.

Seven

Systematic Assessment
for Decisionmaking

Many societal concerns over technical matters arise as diffuse complaints: that doctors are performing too many x-rays or cesarean deliveries, that brown lung disease needs more attention, that television violence is inducing aggressive behavior in children, that housing renewal programs aren't working. Some begin as diffuse senses of opportunity: that wind-power could be tapped, that some exotic wild plants could be domesticated for benefit, that television could be used to educate rural children. Systematic assessment can help shape these concerns into researchable, debatable, resolvable issues.

Prototypical of attempts to anticipate the effects of sociotechnical change was John Evelyn's report of 1664, *Sylva, or A discourse of forest-trees, and the propagation of timber in His Majesties dominions*, which surveyed the "prodigious havoc" being hewn through Britain's forest reserves by the wood demands of shipbuilding, glassmaking, and smelting. The potential of analysis became clear by the mid-1850s. As the astronomer John Frederick William Herschel observed: "Men began to hear with surprise, not unmingled with some vague hope of ultimate benefit, that not only births, deaths, and marriages, but the decisions of tribunals, the results of popular elections, the influence of punishments in checking crime, [and] the comparative value of medical remedies, and . . . different modes of treatment of diseases . . . might come to be surveyed with the lynx-eyed scrutiny of a dispassionate analysis."[1] In *The Coal Question* Stanley Jevons forecasted the winding down of the coal-driven mainspring of England's commerce, and

---

1. Herschel, 1850, p. 12. Shryock, 1961, pp. 230ff., traces early quantitative evaluations of medical practice.

warned his countrymen that they must confront "the momentous choice between brief greatness and longer continued mediocrity" (1865, p. 349). Karl Marx emphasized the relation of analysis to social change in his preface to the first German edition of *Das Kapital* ([1887] 1961, p. 9):

> We should be appalled at the state of things [in Germany] if, as in England, our governments and parliaments appointed periodically commissions of inquiry into economic conditions; if these commissions were armed with the same plenary powers to get at the truth; if it was possible to find for this purpose men as competent, as free from partisanship and respect of persons as are the English factory-inspectors, her medical reporters on public health, her commissioners of inquiry into the exploitation of women and children, into housing and food.

Assessments allow needs, tastes, fears, and hopes to be related to resources, threats, freedoms, and gambles. They allow facts to be sorted out. They allow value transactions to be made explicit. They allow differing appraisals to be compared. They allow uncertainties to be clarified. They allow lay opinions to be compared with those of various experts. And they allow the context of social debates to be revised.

Every segment of industry, government, and public enterprise has to ask of its decisionmaking: How should the necessary scientific and social facts be assembled and evaluated? Are the goals of decision clear, and if not, how can they be clarified? How will consequences, once estimated, be appraised? Can guidance be gained from such methods as cost-effectiveness analysis? How should secondary, indirect, and intangible effects be brought into consideration? How will definitions, analytic boundaries, and conceptual assumptions be established? How should uncertainties be handled? Should formal analyses be depended upon as strict bases for decision, or should they mainly be used as informational background?

This chapter will begin by describing some approaches to analysis; broaden to discuss major studies and reports; describe technology assessment, environmental impact assessment, and social forecasting; survey some meta-analytic considerations; and close by reviewing that archetypal analytic exercise, the Rasmussen Reactor Safety Study.

ANALYTIC APPROACHES

"Prudential Algebra"

Jeremy Bentham explored the "felicific calculus" of decisionmaking; Benjamin Franklin spoke of "prudential algebra"; in personal decisions we "weigh the *pros* and *cons*"; analysts have elaborated these commonsense notions into such methods as decision theory, risk assessment, cost–benefit analysis, and cost-effectiveness analysis.[2]

The very idea of analysis is offensive to some people, as it was to *Saturday Review* editor Norman Cousins (1979, p. 8):

> The world will end neither with a bang nor a whimper but with strident cries of "cost–benefit ratio" by little men with no poetry in their souls. Their measuring sticks will have been meaningless because they are not big enough to be applied to the things that really count.

But in both our private and civic lives we seek benefits, bear costs, and take action on "things that really count" all the time—whether with analytic forethought or not. Why shouldn't things that "count" be counted? I agree with Howard Raiffa:[3]

> We must not pay attention to those voices that say one life is just as precious as 100 lives, or that no amount of money is as important as saving one life. Numbers do count. Such rhetoric leads to emotional, irrational inefficiencies and when life is at stake we should be extremely careful lest we fail to save lives that could have easily been saved with the same resources, or lest we force our disadvantaged poor to spend money that they can ill afford in order to gain a measure of safety that they don't want in comparison to their other more pressing needs.

Analysis usually does *not* consist in asking anything so crass as "How much is one life worth?" but, rather, in asking such questions as "What marginal investment in precaution will preserve marginally how many lives, or person-life-years, or eyes, from being lost to [some] hazard?" or, "How can a given investment protect people most effectively?" Wherever measuring sticks can measure, they should be applied; beyond their reach, obviously not.

Some skeptic is always protesting that the health, environmental, energy, and other decisions I will now describe impossibly amount to "trying to compare apples to oranges"—to which one must respond that life always requires comparing apples to oranges, and to kumquats, ambulances, and aircraft carriers as well. Difficult and frustrating though it may be, it's unavoidable.

Analysis for decision is a problem of handicapping what is likely to happen (the odds of a destructive flood, the probable incidence of a disease, the promise of reading-ability improvement), and appraising quantities

2. Regarding Bentham, see Parekh, 1973. Franklin, letter to Joseph Priestley, September 19, 1772.

3. U.S. House of Representatives, Subcommittee on Science, Research and Technology, 1980, p. 272.

that may be intangible or not easily reducible to currencies of social exchange.

Analytic Factors

Broad decisions often can be factored into more tractable subanalyses of, for example, risks, efficacies, costs, and other characteristics.

Risk is a compound expression of likelihood and severity of deleterious effect. Some risks may be estimated from actuarial records of harm incurred: from government and insurance data we know that in the United States 1,500-some children under thirteen die of automobile injuries every year, and that infants under six months are at highest risk (Baker, 1979; Insurance Institute for Highway Safety, 1981). Or, risks may be estimated from conceptual modeling: to gauge risk from seismic failure of Stone Canyon Dam, which poises a large lake above the University of California at Los Angeles, the probability of earthquakes of given magnitude can be multiplied by the number of people expected to be in the path of the wall of water calculated to be released if the dam ruptures—leading to prediction that there is a 0.0014 chance per year (or more than one chance every thousand years) that an earthquake of magnitude IX will strike Stone Canyon at night, resulting in "substantial chance" that the dam will rupture and instantly kill 207,000 people in the valley (Ayyaswamy et al., 1974).

Risks may be considered to comprise the classical human health afflictions, risks to the environment or to other species, or less classical detriments, such as behavioral stress, addictions, or social disruptions. Risk may be expressed as a differential: "Smokers are 10 times more likely to die from lung cancer than nonsmokers. Heavy smokers are 15 to 25 times more at risk than nonsmokers" (U.S. Surgeon General, 1982, p. vi).

Some risks are, in effect, residual to reduction of other risks. Trace chlorocarbons in municipal drinking water may be generated as minor byproducts of chlorination against harmful bacteria, as trace hydrocarbon contaminants in the feedwater are chlorinated incidentally. Asbestos is a pulmonary hazard, but in many of its uses (fireproof theater curtains, automotive brakes, firewalls) it saves lives. Hexachlorophene sale has been restricted, out of concern over possible toxicity, but hospitals continue to use it, carefully, because of its efficacy as a bactericide. Anesthetics present risks, but they reduce pain and the risks of surgical error, and they make deep operations possible in the first place.

Symmetrically on the other side of the ledger from risk, *efficacy* is the likelihood and intensity of achieving desired effect. As with risk, efficacy can be estimated from experience: the extent to which BCG vaccine protects against tuberculosis, or the success of the Pap test in detecting cervical cancer, or the contraceptive efficacy of a birth control technique. U.S. drug regulations require that prescription drugs be proven efficacious as well as safe. Physicians have used diagnostic efficacy statistics to discriminate more carefully before performing skull x-rays on patients who have suffered head trauma (Bell and Loop, 1971; Phillips, 1979). Current reviews of survival rates and quality-of-life outcomes of coronary artery bypass surgery are helping sort out what kinds of patients are most likely to benefit from the surgery.[4]

In engineering, efficacy may be expressed as component reliability, such as the reliability with which a circuit breaker guards against overload, or as system efficacy at achieving objectives, such as the efficacy of a smoke precipitator in clearing a smokestream. In social research, efficacy is revealed by policy and program evaluations. Laurence Ross's review, *Deterring the Drinking Driver*, examined the record of deterrent efficacy, in several countries, of criminal legal sanctions that differ in modes of enforcement and in severity and certainty of punishment (1982). Both risk and efficacy often can be estimated empirically, even if not precisely, and expressed in quantitative terms.

Benefit can be estimated if the social value of efficacious outcomes can be appraised. The National Academy of Sciences' 1977 study of ionizing radiation struggled with the issue of how to judge the personal and societal benefits of medical x-rays.[5] The Academy's 1979 *Food Safety Policy* report analyzed the benefits of saccharin and of food safety policies regarding mercury, nitrites, and aflatoxin contamination of peanut butter.[6] A committee studying the environmental effects of chlorofluorocarbons drew perspective on the benefits of the chemicals, which are used primarily as refrigerants and spray propellants, by pointing out: "Home refrigeration of food, at one extreme, is important to human health and accounts for less than one percent of all releases. Chlorofluorocarbon uses in aerosol sprays, at the other extreme, are mainly replaceable by other dispensing techniques or by other propellant substances, at some loss in convenience, efficiency, or safety, and amount to about three-quarters of all releases."[7] The 1980 report, *Regulating Pesticides*, described ways of estimating marginal gains in crop yield and consequent benefits from use of pesticides.[8]

Benefits accrue from reduction of risks. Health benefits may be gained, compensation costs saved, materials recycled profitably.

Cost must usually be defined broadly to include not only prosaic market costs but all that has to be paid in materiel and personnel resources, development expenses, commercial and legal transaction fees, insurance costs, and personal freedoms conceded. It is hard to improve on the attitudinal

---

4. U.S. National Institutes of Health, 1981; and U.S. National Center for Health Care Technology, 1981.

5. NAS/NRC, Advisory Committee on the Biological Effects of Ionizing Radiation, 1977.

6. Institute of Medicine/NRC, Committee for a Study on Saccharin and Food Safety Policy, 1979.

7. NAS/NRC, Committee on Impacts of Climatic Change, 1976, pp. 1–7.

8. NAS/NRC, Committee on Prototype Explicit Analyses for Pesticides, 1980.

definition of Thoreau's *Walden*: "The cost of a thing is the amount of what I will call life which is required to be exchanged for it, immediately or in the long run."

Often risks can be translated into costs: hurricane risk to a railway bridge can be appraised as replacement costs and revenue losses. Industrial accidents that knock workers off the job can be costed partially into compensation and substitution expenses, which, though not fully conveying the pain, anguish, and inconvenience incurred, can indicate the job's riskiness and guide hazard-reduction efforts.

Social distribution of risks, costs, and benefits also can be surveyed. This is crucial to valuative discourse. Revision of the Clean Air Act has centered in part around analysis of people's differing exposures and vulnerabilities to pollutants: since the risk to a heavy-smoking Denver asthmatic may be very different from that to a pulmonarily robust Nebraska teenager, what automotive emission levels are "fair"? Analyses of the harm from lead in housepaint, which showed that disadvantaged children living in run-down housing were being poisoned, led to public awareness campaigns, cleanup of old houses, and banning of lead from household paints. In the debate over nitrite meat preservatives in the late 1970s, the fact that poor families are high consumers of cured meats was taken into account when benefits were weighed. Currently, the distribution of sources and effects of acid rain are being analyzed in many countries, preparatory to assigning costs and reducing the pollution. Opening saying, "This report is peculiarly American," the Institute of Medicine's 1981 study of *Health Care in a Context of Civil Rights* investigated "the extent to which race/ethnicity or handicaps affect whether and where people obtain medical care and the quality of that care."[9] Recently the National Center for Health Statistics analyzed the sociodemographic and health characteristics of the fourteen million Americans who have trouble hearing, and found that hearing problems are more prevalent among people having less education and lower family income.[10]

In most such cases, alas, we are much more competent at describing the inequities than we are at deciding what would comprise ethically just redress. Chapter Eight will address the issues of equality and justice.

Weighing the Factors

For decision, the above descriptive-analytic factors must be evaluated together, "trade-offs" made between them, and intangible, political, and equity factors considered.

9. Institute of Medicine, Committee for a Study of the Health Care of Racial/Ethnic Minorities and Handicapped Persons, 1981.

10. U.S. National Center for Health Statistics, 1982.

Risk assessment often is instructive. In risk assessment the health or other risks from various options are appraised, either without consideration for costs and benefits, or operating within assumed cost or benefit givens. At issue may be whether a particular risk is "acceptable," "tolerable," or "as low as is reasonably achievable," or is similar to risks already accepted or to the risks of alternatives.

Most specialists distinguish between *risk analysis* (description), *risk evaluation* (appraisal), and *risk management* (prescription). Usually these can be separated, though they overlap and, as always, their factual and appraisive aspects are interrelated.[11] William Ruckelshaus, when administrator of the Environmental Protection Agency, made a two-part distinction between the fact-finding and prescriptive tasks (1984, p. 157):

> Risk assessment is the use of a base of scientific research to define the probability of some harm coming to an individual or a population as a result of exposure to a substance or situation. Risk management, in contrast, is the public process of deciding what to do where risk has been determined to exist.

When I wrote an earlier book, *Of Acceptable Risk*, I thought it important to try to accommodate the everyday notion, "safety"—after all, major laws in many countries embody the word (1976). I defined "safe" as meaning "of acceptable risk," with "acceptable" carrying a range of denotations, from sprightful gamble (accepting the risks of skiing), to negotiated agreement (accepting the hazards in being a well-paid truckdriver), to passive acquiescence (accepting an inherently dangerous job where there is little personal alternative), to fatalistic stoicism (accepting floods or other "acts of God" as part of one's lot). But now that the regulatory battles of the 1970s have heightened general awareness that risks are relative, that nothing can be absolutely free of risk, and that most risks can be modified, public discussion has quite properly moved away from the word, "safe," toward more contingent interpretations of acceptability. Colloquially it still makes sense to say that a food or a tire is safe; but in more formal contexts qualification ("as safe as any other radial tire") is necessary. The notion of acceptability remains useful. The challenge is to define, for each case, what acceptance means for those affected.

Like the Zen query about the sound of one hand clapping, it doesn't make much sense to ask only about risks without enquiring about benefits and costs. Risk levels that we might tolerate for some benefits, such as protection against fatal disease, we might well hold unacceptable if associated with, say, only cosmetic benefits. And as with many dietary nutrients, mod-

---

11. Lowrance, 1976, Council for Science and Society, 1977, and Royal Society Study Group on Risk, 1983, are general essays. Tobin, 1979, applies the notion of acceptability to sulfur dioxide regulation. Fischhoff et al., 1981, discusses acceptable-risk decisionmaking.

erate doses may be not only beneficial but essential, while high doses may be harmful.

With well-defined projects for which goals and constraints are agreed upon by the major affected parties, for which most of the factors are well known and understood, not only in magnitude but in social distribution, over near and long terms, *risk–benefit* or *cost–benefit accounting* can be useful. As the names imply, these approaches tally up pluses and minuses, assign pricetags where they can, and sum net gains or losses incurred by manufacturers, government programs, or other affected parties or "society." Attributes that cannot be reduced to semiquantitative terms, such as freedoms or fairness or anguish, can be brought along as ancillary (not necessarily unimportant) factors. The occasional determinative application of such techniques—and their labels' rational ring—tempts legislators, administrators, managers, judges, and presidents to call for their use. Because the qualifying conditions I described above rarely obtain, the approach is more useful for broad estimation than for precise appraisal.[12]

Cost-effectiveness analysis compares the effectiveness with which various alternative actions can be expected to achieve stated objectives at given expenditure: how automobile seat belts compare with other forms of passive restraint in limiting injury, how kidney transplants compare to dialysis in compensating for renal failure. In choosing among medical screening and therapeutic options, cost-effectiveness analysis has proven extremely useful.[13] An extreme challenge to the method is the question, "Does psychotherapy provide sufficient return to warrant being reimbursed under Medicare?" being asked by such powerful billpayers as the Senate Finance Committee.[14]

Decision analysis, an offshoot of operations research, lays out a decision tree of possible events and options, assigns probabilities to each of the branches and values to each of the final outcomes, then compounds backward to appraise the decisional choices. The method is particularly useful for analyzing technical programs designed to cope with natural hazards, such as hurricanes, floods, or well-known epidemic diseases, which recur and are susceptible of probabilistic analysis. Probabilities and utilities still have to be estimated, which, as with other analytic methods, resolves to "voting" over facts and values by the analysts and those whose opinions they canvass.[15]

12. Bunker, Barnes, and Mosteller, 1977, analyzes costs, risks, and benefits of surgery. Haimes, 1981, addresses risk–benefit analysis of water resources. Swartzman, Liroff, and Croke, 1982, discusses cost–benefit analysis in environmental regulation.

13. Eddy, 1980; Berwick, 1980; and U.S. Congress, Office of Technology Assessment, 1978, August 1980.

14. U.S. Congress, Office of Technology Assessment, October 1980.

15. Keeney and Raiffa, 1976, is a hard-slogging classic. Stokey and Zeckhauser, 1978, is a readable introduction. Weinstein and Fineberg, 1980, is a fine text on clinical decision analysis.

Decision-analytic techniques proved very useful for analyzing what to do about perchloroethylene (PCE), the commercial drycleaning solvent. Analysts developed a quantitative description of the exposures and cancer risks PCE poses to workers and others, examined the marginal reduction in cancer risks that could be achieved by various mechanical and procedural options, and then, after assigning explicit worth to cases-of-cancer-avoided, determined which control strategies would yield optimal health protection (Campbell, Cohan, and North, 1982). Similarly, Milton Weinstein put forth a provocative example by analyzing the cancer-reduction "payoff" of large-scale rodent carcinogenicity assays, then comparing this to the possible returns from investigation of beta-carotene as a dietary cancer-prevention agent.[16]

LIMITATIONS OF, AND CHALLENGES TO, ANALYSIS

Limitations of Analysis

The shortcomings of formal analysis are many. I am convinced that these approaches are useful, indeed almost necessary, for structuring discussion and informing decisionmakers, but that they are less useful and more abusable when bent to narrowly legalistic purposes. In their *Primer for Policy Analysis* Edith Stokey and Richard Zeckhauser correctly warned (1978, p. 135):

> Benefit–cost analysis is especially vulnerable to misapplication through carelessness, naïveté, or outright deception. The techniques are potentially dangerous to the extent that they convey an aura of precision and objectivity. Logically they can be no more precise than the assumptions and valuations that they employ; frequently, through the compounding of errors, they may be less so. Deception is quite a different matter, involving submerged assumptions, unfairly chosen valuations, and purposeful misestimates. Bureaucratic agencies, for example, have powerful incentives to underestimate the costs of proposed projects. Any procedure for making policy choices, from divine guidance to computer algorithms, can be manipulated unfairly.

Stokey and Zeckhauser emphasized that "prudential algebra" of one form or another must be resorted to nevertheless, and in their text illustrated many applications.

All analytical approaches have difficulty with scientific uncertainties, with envisioning all possible consequences, with placing "prices" on preserving human life and environmental goods, with taking into account

16. Weinstein, 1983, and letters of commentary appearing in *Science* 222, pp. 1072–1075, 1983.

intangibles and amenities in general, with assessing the social costs of opportunities forgone, and with anticipating and "being fair" to the future. Some crucial but hardly predictable factors—such as weapons proliferation and terrorist risks from the spread of civilian nuclear power, or the amenity value of national parklands to future generations—almost elude accounting.

Challenges to Analysis

The first challenge is to develop analytic approaches that directly inform, illuminate, and aid decisionmaking that affects the public. Although we can't expect perfectly "rational" public decisions, we must continue to try to develop rational perspective, including perspective on our irrationalities.

Regulatory decisionmaking heavily depends on analysis. Under Section 6 of the Toxic Substance Control Act the Environmental Protection Agency must protect the public against "unreasonable risk of injury"; under the Section 112 stationary-sources provisions of the Clean Air Act it must ensure "an ample margin of safety to protect the public health"; under Section 1412 of the Safe Drinking Water Act it must protect the public "to the extent feasible . . . (taking costs into consideration)." In these laws "unreasonable risk," "ample margin of safety," and "feasible" are left undefined. Hence analysis—of risks, costs, benefits, equities—is at the essence of the Agency's determinations.

In this regard, we are going to have to admit that most legislated decision frameworks are, really, only versions of one framework. Regulation drafters often distinguish between "absolute no-risk" rules, benefit–risk balancing, cost-effectiveness, "best available technological control," protection "as low as is reasonably achievable," and the like. But in actual implementation, these amount to distinctions without differences: zero-risk rules have simply been ignored when obeying them would be too costly for the benefit achieved; "available" control and "reasonable" achievability have been defined, pragmatically, by benefit–cost considerations; and although cost-effectiveness analysis does not estimate benefits, it assumes the benefits to be worth the effectively spent costs.

In regulation, a key question is how much formal legal status to accord analyses. This was at the heart of the debate in the late 1970s over the Occupational Safety and Health Administration's regulation of benzene in gasoline. The Supreme Court's less-than-conclusive decision in *Industrial Union Department, AFL-CIO* v. *American Petroleum Institute* affirmed a lower court's ruling that "OSHA had exceeded its standard-setting authority because it had not been shown that the one-part-per-million exposure limit was 'reasonably necessary or appropriate to provide safe and healthful employment' as required by [the Occupational Safety and Health Act's] Section 3(8), and that [the Act's] Section 6(b)(5) did not give OSHA the unbridled discretion to adopt standards designed to create absolutely risk-free workplaces regardless of cost."[17] Ancillary to this was the question of whether analysis could patently dictate the OSHA Administrator's decision, beyond merely informing him of estimates. This issue remains unresolved. I think analyses should be used as fully, and be pressed as far, as possible, but that public administrators should be left free to make their decisions on a variety of bases, using analyses as foci of discussion.[18]

A correlate challenge to analysis is to develop explicit and comparative approaches. My view is that we should estimate what costs really will add up to, and what returns investments will buy, and who will benefit and who bear burdens.[19] We now undertake to intercompare various energy options, and earthquake-hazard-reduction options, food sweetener options, and airport siting options. Usually it is important to examine entire systems. Crosscutting appraisals provide essential perspective among programs. The next chapter will describe ways life-preserving investments are appraised.

Another challenge for analysis is to draw perspective on its results. David Eddy's introduction to his analysis of breast cancer screening dramatically raised this problem (1980, p. 25):

> Suppose that a certain screening program can be expected to decrease by 25% the chance that a woman will die of cancer of the breast over the next five years, or increase her life expectancy by 80 days, at a cost of $400. Now what? There remains the problem of making tradeoffs. It is not enough to say that we want to maximize life expectancy; that objective leads to the decision to do universal prophylactic mastectomies. Nor is it enough to say that we want to minimize costs; that implies that we padlock the hospital doors. If a program increases the chance that a woman gets cancer [from x-ray examination], but decreases the chance that she dies of it, we must compare the increased anxiety and costs of having cancer with the pleasure derived from living longer. Which is preferable: program A that increases life expectancy 50 days and costs $100, or program B that increases life expectancy 80 days and costs $400? How should we even talk about the benefits or effectiveness of a screening program? Life expectancy? The chance of dying in 5 years? The chance of dying of breast cancer in 5 years? Ten years? The chance of having a cancer detected at a screening session? Clearly, these questions are difficult, yet they are unavoidable.

Eddy went on to show how, when placed in broad, comparative, appraisive context, such analytic estimates can inform decisionmaking by women, their doctors, and the medical industry.

---

17. *Industrial Union Department, AFL-CIO* v. *American Petroleum Institute*, 448 U.S. 607 (1980).

18. For commentary see DiSanti, 1980, and Bartman, 1982.

19. Lowrance, 1982a, discusses examples.

MAJOR STUDIES AND REPORTS

Present times can well be called the Age of Studies. In dealing with complex, uncertain, value-laden issues, society's reconnoiterings have become much more formal in organization, broader in scope, and more collective in preparation. Sociotechnical reports having implications for industrial, regulatory, and public policy decisions are turned out at an overwhelming rate. Usually these studies are critical consolidations of the research literature and of analyses like those discussed in the preceding section. Society has come to depend on them.

Among the first assessments commissioned for public inquiry were the steamboiler investigations of the early 1800s. Pressurized boilers were being fabricated from soft copper and brittle cast iron instead of tougher wrought iron. Steam relief valves were being abused, as by being tied down during riverboat races. Poorly trained firemen were operating the machinery without understanding how it worked. As a result, hundreds of people were killed by exploding boilers. The newly formed Franklin Institute of Philadelphia performed experiments, conducted inquiries of accidents, and published findings on a whole range of questions about boiler design, construction, operation, and maintenance. Both in analytic approach and in reportorial tone the Institute's 1836 *Report on the Explosions of Steam Boilers* set lasting precedent. Also prefiguring current woes, politicians largely ignored the report's recommendations for years, until catastrophic explosions occurred (Burke, 1966; Sinclair, 1966).

Since then, in pursuit of breadth and balance, benchmark reviews increasingly have been prepared by committees. I would recall such gathering points as the President's Science Advisory Committee reports on *Seismic Improvement* [related to nuclear explosion verification] (1959), on *The World Food Problem* (1967), and on the herbicide *2,4,5-T* (1971);[20] the Coleman report on *Equality of Educational Opportunity* for the Secretary of Health, Education, and Welfare (1966);[21] the *Report of the [HEW] Secretary's Commission on Pesticides and Their Relationship to Environmental Health* (1969);[22] the National Academy of Sciences' *Effects on Populations of Exposure to Low Levels of Ionizing Radiation* (1972) (understandably, better known as the BEIR—biological effects of ionizing radiation—report), its saccharin-triggered *Food Safety Policy* report (1978–1979), and *Health Care for American Veterans* (1977);[23] the Nuclear Regulatory Commission's (Ras-

20. U.S. President's Science Advisory Committee, 1959, 1967, 1971. PSAC's work is recounted by Beckler, 1974.

21. Coleman, 1966. Harvard Educational Review, 1969, provides commentary.

22. U.S. Department of Health, Education, and Welfare, 1969.

23. NAS/NRC, Advisory Committee on the Biological Effects of Ionizing Radiations, 1972. Institute of Medicine/NRC, Committee for a Study on Saccharin and Food Safety Policy, 1979. NRC, Committee on Health Care Resources in the Veterans Administration, 1977.

mussen) *Reactor Safety Study* (1975), which I will describe shortly;[24] the Defense Science Board's (Bucy) *Analysis of Export Control of U.S. Technology* (1976);[25] and the *Report of the President's Commission on the Accident at Three Mile Island* (1979a), whose complaints about design, operation, and technical management, incidentally, echoed many of those nineteenth-century boiler-accident inquests.[26] The effects of all of these are still being felt.

Thousands of other reports have been prepared besides these center-pieces. Few major sociotechnical actions now are undertaken without prior assessment. Inquiries and public reports are expected whenever a dam breaks, an airliner crashes, a wildlife habitat comes under strain, or a public health threat arises.

Such studies are part of science's cumulation and filtering process in that they critique existing primary literature, reviews, monographs, symposia, and other reports. Usually they dovetail technical considerations into broader societal issues. And they provide feedback to the R&D agenda. In preparing its 1982 study, *Television and Behavior*, the National Institute of Mental Health and its advisors reviewed some 2,500 primary research reports published during the preceding ten years, discussed such related issues as television's effects on social life and education, and, although it did not make recommendations, raised a wide variety of research and policy considerations (U.S. National Institute of Mental Health, 1982).

Serious studies are tedious to prepare. Most are only mediocre. Many are badly flawed. It is easy to question whether they are useful; many, of course, are not. Overall, though, I must agree with Philip Handler, who, looking back over his experience with hundreds of studies while president of the National Academy of Sciences, reflected (1978, p. 9):

> It is commonplace to scoff at "committees," to suggest that the most useful committee has but one member, to propose that some alternative mechanism should be utilized for the analysis and resolution of questions. . . . It has been suggested that formal adversarial procedures would be more suitable than the committee system for the examination and analysis of technical matters that are the subject of public controversy. But neither the hearing procedures of legislative bodies nor the formal procedures of administrative law judges, of hearing commissioners, or of the court system have been demonstrated to be as effective as a well-constructed committee for the analysis and resolution of complex controversy regarding technical matters.

---

24. U.S. Nuclear Regulatory Commission, 1975.

25. U.S. Department of Defense, Defense Science Board, 1976.

26. U.S. President's Commission on the Accident at Three Mile Island, 1979.

The Craft of Study Design

The social legitimacy and usefulness of reports depend not just on the documents' internal logic but, more strongly than many preparers have assumed, on the dynamics of conception and preparation.

CHARTER AND BOUNDARIES. Since major studies are addressed to decision-making clienteles (in effect if not always explicitly), it is crucial that report preparers negotiate terms of reference with their patrons and potential users to ensure that their report will address apposite questions. The charter must be explicit and clear. Too many reports go directly from printing press to office bookshelves without being read because they simply don't ask the pertinent questions. Committees can benefit from laying out, early in their project, an array of possible findings and asking of each, So what? How would having this information affect the decisions at issue?

Obviously the way a study is defined will affect its outcome. From what perspectives will the issues be examined? What are the boundaries of inquiry? Are any topics off-limits? How technical should the report be? How will it handle social value judgments? Should it address policy questions specifically?

SELECTION OF PREPARERS. The only way a person can be perfectly unbiased is to be perfectly uninformed. The intention in appointing a technical committee should be to seek flexible, fair members having a diversity of skills and interests, not unbiasedness. Deep experts can be drawn from varied backgrounds. Nontechnical and technical generalists can be included. Experts from disciplines intellectually adjacent to those central to the matter under study can play critical roles. Representativeness can be sought by including or consulting people from a wide variety of disciplinary, institutional, and sociopolitical backgrounds.

Some institutions require that study preparers file on record a statement regarding their potential for financial, political, or personal conflicts of interest. They may also ask new committee members to describe any relevant public stands they have taken on the issues, and to reveal any circumstances that would compromise their professional independence in rendering opinion. Public candor can be offered by placing these statements on open record, not just keeping them confidential within the institution.

HANDLING OF META-ANALYTIC ISSUES. Committees should be instructed to tell the truth, the whole truth, nothing but the truth, and the uncertainty about the truth. They should be urged to critique their assumptions and analytic approaches, address value-laden judgments explicitly and candidly, reveal the limits of their confidence in the findings, indicate how dependent the conclusions are on the assumptions and methods employed, and identify uncertainties that can be reduced by further research or assessment.

REPRESENTATIVENESS OF ESTIMATES. Studies always have to make judgments as to whether to use median estimates (of costs, say, or risks) or extreme estimates. Addressing this issue, President Reagan's Executive Order No. 12,291 of February 1981 directed that regulatory analyses be based on "most likely" or "best" estimates, and not on "conservative" (extremely cautious) or "worst-case" estimates; bureaucrats and advisors are still trying to define what that means. Engineers always have added somewhat arbitrary "safety factors" onto design specifications; but now, cramped on costs, they are tending to calculate what overstresses could possibly be experienced and to design against those possible extremes but not much further. Scientists analyzing allergins, floods, riots, epidemics, and other disruptions face similar difficulties. To what extent to protect especially vulnerable people? How fully to anticipate very unlikely events? How to weight frequent but small effects compared to large but very infrequent ones? Sophisticated analyses assess *ranges* of possible effects.

CONSENSUS? Panels usually try to reach consensus. This may not be possible on all aspects of a study. On matters not reaching consensus, the report may publish opinions dissenting from the majority's. In context of ongoing debate it is very important that reports reveal the push-and-pull of opinion, axes of disagreement, and reasons for rejecting alternative conclusions and recommendations.

INSTITUTIONAL IMPRIMATUR. Organizations sometimes use the tactic of releasing internal staff or consultants' reports as "lightning-rod" documents to draw early reaction, to be disowned by the institution if necessary. But institutional imprimatur is almost always sought for major studies, as legitimation. Usually such reports must be approved by a board of directors acting on behalf of the organization.

Confusion, for the preparing body and for the public or other clients, arises when report preparers and their host institution can't reach resolution. Such a problem arose, for example, in 1982 when Frank Press, as chairman of the National Research Council, transmitted an NRC committee's *Analysis of Marijuana Policy* to the National Institute on Drug Abuse with evident reluctance: in his preface to the report Press asserted that "there is no new scientific information exonerating marijuana," and said, "The only position that can be inferred with respect to the National Research Council on the issue of marijuana policy is that the National Research Council is satisfied that the Committee was competent to examine the issue and diligent in carrying out its task." Faint imprimatur indeed. I suspect the disagreement hinged both on what might constitute "new [social-] scientific information" about effects of decriminalization, and on whether the committee's charter directed it to choose among and endorse particular social options.[27]

RELATION TO PATRONS AND CLIENTS. From the beginning it is crucial for a study project to clarify how it will relate to those who will pay for and publish the study, to those who will be consulted during the study, to potential users of the report, and to those who will be affected by the study. The difficulty for the committee is to stay receptive enough to learn the complexion and details of the problems and tailor the study to decisional needs, while keeping distant enough to remain, and preserve the appearance of remaining, insulated from undue influence. The institutional context, the financial grant or contract, and the terms of reference for the study are key. Issues include whether sponsors or clients, such as an industry or government agency, should participate in the meetings and review the report in draft, and whether the report should make recommendations and draw out the societal implications of the technical findings.

PRESENTATION AND RELEASE OF FINDINGS. Advice to write clearly may seem unnecessary; but the prolixity of orotund, jargonized documents that languish in passive storage-mode noninterfaced with the real world attests otherwise. Too, scientists, who demand terse, informative summaries in scientific research reports, often turn around and deliver to busy officials plump documents carrying bland summaries and weakly focused recommendations. Yet, especially at higher levels of management and government, these may well be the only parts of reports that get read closely. Committees should be urged to combat the fatigue factor by setting reasonable deadlines and by reserving a few especially articulate committee members to write the summarizing material and abridged "executive" versions of their report in the last days before deadline.

All conclusions should be substantiated by evidence. Again, this should be unnecessary advice but is not. The National Academy's Food and Nutrition Board violated this precept in 1980 when it published a skinny report, *Toward Healthful Diets*, that had the effect of contradicting a wide range of prevailing scientific opinions about, among other things, the effects of cholesterol intake on heart disease. But the report did not publish evidence in support of its claim, and did not demonstrate why its opinion should supplant current government and Academy advice.[28] The Board's findings, which essentially were just a plea for dietary moderation, may well prove

27. NRC, Committee on Substance Abuse and Habitual Behavior, 1982. A follow-up statement by the committee chairman, Louis Lasagna, appeared in *The Sciences* 22, no. 9, pp. 6–7, December 1982.

28. NAS/NRC, Food and Nutrition Board, 1980.

correct. But for many reasons, in part for not publishing a complete report, the Board came under heavy criticism that undid any good its report might have served.[29]

Many a slip can occur between text and summary. A recent monograph, *Clipped Wings: The American SST Conflict*, recounted the saga of the preparation of the Department of Transportation's 7,200-page supersonic transport (SST) impact assessment.[30]

> At a press conference, the Department distributed a twenty-seven-page summary of the whole report that emphasized the report's optimistic conclusions that the currently planned total of 30 Concordes and TU-l44s presented no appreciable stratospheric hazard and that there was a high likelihood that low-emission engines and fuels would be developed in the near future. The summary ignored the effect of a large fleet of SSTs. The press mistakenly reported that the director of this study had stated that a full-scale American SST fleet would not weaken the ozone shield. "Scientists clear the SST," the *Christian Science Monitor* declared.

The executive summary of the original Rasmussen *Reactor Safety Study* was judged to be such a poor digest of the full report that it had to be officially withdrawn, adding confusion to a controversy that was already too confusing.

Preliminary release of reports can open them to attack when they are only tentatively formed. On the other hand, releasing a complete draft for comment, as many government agencies do, can attract suggestions that greatly enrich the final publication. Policy on prepublication review and release should be decided upon early along in projects.

FOLLOW-THROUGH. A key question for technical committees is how vigorously to follow their newly printed report into the world of action. Press conferences usually help gain reasonable coverage of the document. Briefing sessions can be held. Beyond that, hearings or symposia can be convened to discuss the report. Sometimes committee members personally convey their report to public leaders. In the flood of documents crossing officials' desks, a report that is not central to the recipients' duties and demanding of response may, if not bolstered by publicity or endorsement, well get ignored.

An extreme and effective kind of follow-through is preparation of an update report, as when the National Research Council in 1982 published

29. For hearings on the report, which congressional committee members casted as a morality play, see U.S. House of Representatives, Subcommittee on Domestic Marketing, Consumer Relations, and Nutrition, of the Committee on Agriculture, 1980; and Wade, 1980.

30. Horwitch, 1982, pp. 336–338. Also see U.S. House of Representatives, Government Activities and Transportation Subcommittee, 1975.

Causes and Effects of Stratospheric Ozone Reduction: An Update to reassess that controversy several years after the NRC's earlier studies.[31]

Critical Review of Studies

Rarely are studies subjected to the broad, serious, constructively critical review they warrant. Most reports are reviewed by the originating institution before they are released. Directly affected groups discuss the findings. But debate over uncertainties, assumptions, social value judgments, context of controversy, and usefulness to decisionmaking is not often carried effectively into larger circles. The documents may not even be readily accessible. Regardless of how balanced reports may be, intellectual or political antagonists are more likely to take up the issues than middle-ground observers and synthesizers are.

Recent studies I would like to have seen explicated include, as examples from a long list, the study *Risks of Energy Production* that Herbert Inhaber prepared for the Canadian Atomic Energy Control Board in 1978, the Council on Environmental Quality's *Global 2000 Report*, and Richard Doll and Richard Peto's report to the Office of Technology Assessment on avoidable cancer risks.[32] In all three studies the scientific bases were uncertain, and the conclusions had broad social implications.

Reports that are solid should be endorsed; those not, struck down. "Criticism" need not carry pejorative overtones. We urgently need to develop ways for conducting broadly exposed review, which can have important buffering and leavening effect. I have proposed that *Science*, the *New England Journal of Medicine*, and other journals solicit in-depth essays on recently released reports that have central relevance to public decisions (1982b). These essays would amplify the reports, interpret them for a broader readership, critique their assumptions, methods, and findings, and place them in perspective relative to other studies and to current events. Such review would manifest scientific community in its original sense, provide partial answer to "Who will watch the watchers?" and offer much-needed social stewardship. Reports do come under public comment, of course. My concern is that we institute ways for fostering constructive review as a regular activity.

Study Usefulness

Are studies useful? Pointless here to try to generalize. Some are, some aren't. Studies are used to fill information gaps, help groups assert their

---

31. NAS/NRC, Committee on Chemistry and Physics of Ozone Depletion and Committee on Biological Effects of Increased Solar Ultraviolet Radiation, 1982.

32. Inhaber, 1982, is a reprint of the revised report and critics' reviews. U.S. Council on Environmental Quality and Department of State, 1980. Doll and Peto, 1981.

roles and express their views, buttress or refute preexisting arguments, and build a record. Participants often emphasize that the process of hammering out reports serves as much use as the final printed documents do—pulling into juxtaposition differing perspectives, intuitions, assumptions, weightings, and apprehensions that might not otherwise come into comparison, and reducing communication barriers among the participants. Over disturbing technical issues of wide import, society demands high-level review. In the aftermath of Three Mile Island, Americans simply would not have tolerated going without an investigation like that by the President's Commission.

Legitimacy and authoritativeness are key. Correctness is essential, but reports that are prepared and endorsed by reputable individuals and institutions and that respect procedural proprieties "go further" than reports that aren't and don't. Major studies are a formal version of peer review.

Studies cumulate: current assessments of nuclear power depend on studies built on the revised Rasmussen reactor study, which was built on many prior reactor evaluations and on such studies as the BEIR report on health effects of radiation, which in turn had drawn heavily on such studies as those of the Atomic Bomb Casualty Commission. . . .

"Perhaps the cardinal attribute of the 'successful' report is felicitous timing," Philip Handler remarked (1979, p. 4).

> Even a relatively poorly prepared report that is placed in the public domain at the right historic moment, when, in some sense, the stage has already been set, can occasionally find immediate acceptance and initiate events of historic importance. Conversely, an excellent report released into an unprepared environment, when the times are not propitious, is unlikely to alter the course of human affairs. Regardless of how compelling its case, if too far in advance of its time, it may seem to have no impact. Only later will the report be recognized to have been part of the stage-setting process.

Timing need not be passive, but can be influenced by proper laying of groundwork, keying of the study to evident needs, and involving in the planning those who will be affected by the report.

Solid reports on contentious issues can be disputed but can hardly be ignored. Reports on issues that have not yet gained society's attention can more easily be overlooked. Regardless of their original reception, for a long time after publication thorough reports serve as conceptual cairns, being both landmarks and troves of information.

ANTICIPATORY ASSESSMENT

In 1946 William Ogburn published a sweeping survey and forecast, *The Social Effects of Aviation*, which anticipated the new technology's effects on demographics, health, agriculture, and many other aspects of life (1946). It was a fascinating book, and almost all of its prophecies have been realized. Moreover, the project demonstrated that such crystal-gazing was both feasible and useful.

Briefly, now, we should take note of the systematic approaches that have been developing since then: technology assessment, environmental impact assessment, and social forecasting.

Technology assessment became a defined endeavor—some would even call it a movement—in the late 1960s. Its ambitious purpose is to assess the primary, secondary, and derivative consequences of emerging or expanding technologies, including their tangible and intangible effects, over the intermediate to long term, on health, employment, communication, or other aspects of life. It can involve a variety of methods: trend projection, polling of expert opinions, cost–benefit analysis. The intention is to describe, forecast, and possibly to evaluate for decisionmaking.

In 1972 the Congress established an Office of Technology Assessment (OTA) "to provide early indications of the probable beneficial and adverse impacts of the applications of technology" and develop related information. OTA has made some important technical advisory contributions, but it has not yet subjected many major technological innovations to full-scale anticipatory scrutiny. Most of its reports, such as *The Future of Natural Gas Imports* and *Physician Supply and Requirements*, are simply technical advice. Operating in the congressional arena, under a bipartisan, bicameral board, with limited resources, rarely is OTA assigned critical review of truly innovative technologies. As OTA's report titles indicate—*Coal Slurry Pipelines*, *Policy Implications of the Computed Tomography (CT) Scanner*—it tends to work on problems of intermediate technical novelty and complexity and less-than-enormous societal impact. However, sometimes OTA examines emerging technologies that are not clearly included in other bureaus' portfolios, as it did in *Scientific Validity of Polygraph [lie detector] Testing*. As with its study of *MX Missile Basing*, it can provide the Congress with assessments of administrative branch proposals. The OTA has potential, and it should be encouraged and strengthened.[33]

Many of the studies referred to earlier in this chapter could be considered technology assessments. So could some regulatory evaluations, such as those of pharmaceuticals. So could some of the assessments of recombinant DNA. And so could some national commission studies, such as that by the National Commission on Electronic Funds Transfer.[34]

---

33. For general commentary and bibliography on the OTA, see U.S. Congress, Congressional Research Service, 1979, pp. 831–987. O'Brien and Marchand, 1982, carries recent essays on technology assessment.

34. U.S. National Commission on Electronic Funds Transfer, 1977. For commentary, see Colton and Kraemer, 1982.

There have been a few excellent academic technology assessments, such as Edward Dickson and Raymond Bowers's study of the video telephone (1974). Also, there have been a few "retrospective" technology assessments, "historical case studies of the development of a new technology and the consequences of that development for society"; a group at the George Washington University conducted an exemplary one, *Submarine Telegraphy* (Coates and Finn, 1979).

Environmental impact assessment has related purposes. These have most notably been embodied in the National Environmental Policy Act (NEPA) of 1969. This procedural law requires that federal agencies file statements with the Council on Environmental Quality on the environmental impacts of their anticipated actions, alternatives to the proposed actions, and irreversible effects. As with technology assessment, it can use a variety of analytic methods.

"NEPA may be seen as a contrived, institutionalized answer to a people's recognition of its own deficiencies," Lynton Caldwell has said. "Through the impact assessment process . . . we compel ourselves . . . to do what we know should be done in undertaking actions that may have consequences not immediately apparent. The environmental impact statement process institutionalizes patience, caution, and looking before leaping."[35]

Social forecasting is a related approach. In 1933, President Hoover's Research Committee on Social Trends prepared the ambitious and moderately influential report *Recent Social Trends in the United States*, which took stock of changes in communication, health, employment, energy supply, family structure, and other aspects of American life.[36] A recent successor was *The Global 2000 Report*, commissioned by President Carter to study the "probable changes in the world's population, natural resources, and environment through the end of the century," meant to serve as the foundation for longer-term planning.[37] Many other such efforts can be recalled, such as *Limits to Growth* and other population and resources projections prepared by the Club of Rome, and the medium-term energy assessments a number of organizations conducted during the 1970s.[38]

The National Academy of Sciences' ambitious study, *Energy in Transition, 1985–2010*, illustrated how dependent such complex forecasts are on

---

35. Caldwell, 1982, p. 150; this book provides an overview of the literature on environmental impact assessment. Holling, 1978, is a standard text.

36. U.S. President's Research Committee on Social Trends, 1933. For commentary, see Karl, 1974.

37. U.S. Council on Environmental Quality and Department of State, 1980. Simon and Kahn, 1984, is a refutation.

38. Ascher, 1978, is an excellent review of the problems of technical forecasting. Cahn and Cahn, 1980, is a good history. The exchange between Julian Simon, 1980, and respondants in *Science* 210, pp. 1296–1305, 1980, illustrates many problems of global prediction. Greenberger, 1983, offers critical commentary on fourteen American energy studies.

their assumptions, and how vulnerable they are to social discontinuities. That study, undertaken in 1975 as "a detailed and objective analysis of the risks and benefits associated with alternative conventional and breeder reactors as sources of power," looked toward the period 1985–2010. But as Philip Handler's letter of transmittal emphasized, many changes occurred between 1975 and 1980 that influenced the projections:[39]

> There were gasoline shortages and price rises, electricity blackouts, natural gas shortages, public debate over power plant sitings, large negative balances of payments for petroleum and for technology. Growing environmental concern was paralleled by concern that regulation is inhibiting industrial innovation and productivity. Rising prices and the debate over decontrol were accompanied by growing public distrust of the energy industries and of statements concerning the magnitude of hydrocarbon reserves. Political instability in nations on which we depend for petroleum imports made all too obvious the precariousness of the flow of imported oil. Three Mile Island revealed both the resilience designed into nuclear plants and the significance of the human factor in the operation of such plants. Established energy companies began to develop capabilities in new energy technologies, and a host of new, smaller companies entered the market for such technologies as solar heating, windmills, biomass utilization, insulation, etc.
>
> President Carter, particularly concerned that nuclear weapons should not proliferate, took action to defer reprocessing of spent nuclear materials and to delay commercialization of a breeder reactor, while the pace of the much debated Clinch River breeder project was deliberately slowed.

Need anyone wonder why the report's projections to 2010 were controversial, or why they remain tenuous?

For any technology assessment, environmental impact assessment, or social forecast, the boundary "scoping" questions are: Are the time horizons being projected toward appropriate? Are the principal future alternatives properly envisioned? Are the modes of impact being surveyed the most important ones? Are social values being taken into account appropriately? And is the assessment effectively oriented to decisionmaking?

META-ANALYTIC CONSIDERATIONS

All assessments are deeply affected by underlying, or flanking, meta-analytic considerations: selection of analytic approaches, adoption of working

---

39. NAS/NRC, Committee on Nuclear and Alternative Energy Systems, 1980, p. vii.

assumptions, fact/value integration, and handling of factual uncertainty and indeterminacy. Because often they are less-than-consciously presumed and less-than-fully recognized and articulated, merely to call them "assumptions" (like mathematical preconditions) is insufficient. Too, we need a term that suggests need for constant critical reexamination. The Greek prefix *"meta-"*—"foundational; standing beside or behind; ulterior to; bounding"—as used in "meta-theological" and "meta-mathematical," is the apt modifier. Thus, "meta-analytic."

At this point more needs to be said about fact/value integration and about the handling of factual uncertainty and indeterminacy.

The Fact/Value Distinction, Again

Repeatedly I have urged that matters of technical fact be kept distinct from matters of personal and social value judgment, to the greatest extent possible. Yet I have expressed skepticism about the ideal of "objectivity," and have argued that all technical research, professional practice, assessment, and advising is value-freighted. This need not be seen as constituting a quandary, although it raises difficult practical problems. What are wrong are the two absolutist views: at one extreme that all fact-finding can be pristinely objective and value-free; at the other that since all facts are merely subjective personal opinion, all views are equivalent as claims to truth.

At any given time in a given social group, large bodies of technical understanding are held as solid, confirmed, noncontroversial, useful, or defined-as-true. Some knowledge may be held certain except in rare domains of experience, such as in outer space or under high pressure. This corpus is embraced strongly and remains independent of social passions, though it is always subject to change. (The speed of light in a vacuum is understood everywhere in the world to be exactly 299,792,458 meters per second.) Likewise, at any point in history, in a given social group, aspirations and preferences are guided by a large body of consensus values, implicit and explicit. (Leprosy is abhorred everywhere.) Within any group, despite uncertainties and disagreements, slow-changing core values dominate. In appraisals, therefore, many factors—even though they are value-laden—can be adopted as givens. People can't and don't go back to philosophic foundations on every aspect of every decision, but debate selected issues.

Factual questioning is conditioned by valuative convictions about *which things are important, what is at stake*. Value inquiries are predicated on facts as currently understood about *what was, what is, what will be*, and *what can be*. Special contention arises where somewhat uncertain science intersects with somewhat uncertain values.

It is easy to see how an assessment can hybridize factual and evaluative considerations. For example, five members of the thirty-seven-person

National Academy of Sciences' Committee on Saccharin and Food Safety Policy committee appended a minority statement to the group's report arguing, among other things, that:

> Classification of food additives into risk categories such as high or moderate for regulatory purposes cannot be done using current scientific data and theories; . . . irreversible toxicities are deserving of special regulations; . . . risks from foods should be lower than other types of risks; . . . direct food additives should be regulated differently than other classes of food additives or contaminants; and . . . the classification of saccharin and its regulatory status are unacceptable.

In transmitting the full report Academy president Philip Handler explained: "The difference of opinion which led to this ambivalent statement is not a differing interpretation of scientific fact or observation; it reflects, rather, seriously differing value systems."[40]

In policy analyses the problem takes extreme form, as Ruth Hanft made clear about health-care policy (1983, p. 254):

> Most social scientists who propose a competitive model to address the cost problems in health care assume that a free market approach will work in health care, that supply and demand will reach an equilibrium, that prices will respond to the actions and reactions of supply and demand, and that goods will be distributed equitably. This model also assumes the following conditions:
>
> 1. Consumers have enough information to make rational choices.
> 2. Suppliers have free entry into the economic market.
> 3. Most demand is created by consumers, and demand can be withheld or delayed.
> 4. Demand induced by providers can be reduced through economic incentives.
> 5. Prices will fall if demand falls or supply increases.
> 6. If consumers pay directly for services (rather than through third-party payers) they will act as rational purchasers—shopping for the best buy.
> 7. Supply will expand or contract in relation to demand.
>
> The regulation school, on the other hand, assumes that a medical market cannot operate as a free economic market or ensure equitable distribution of services for the following reasons:
>
> 1. Consumers can never have sufficient technical information to make truly informed choices.

---

40. Institute of Medicine/NRC, Committee for a Study on Saccharin and Food Safety Policy, 1979.

2. There is no free entry of suppliers into the economic market because of licensure and other constraints related to quality.
3. The life, death, and disability results of choice, combined with the need for highly technical information, require that the consumer have a representative—the physician.
4. Demand is often created by the agent, who has an economic stake in providing services.
5. Direct payment at time of use may influence demand marginally, but it has less than normal influence when the product is related to urgent health needs or pain.
6. Direct payment at time of use acts as a barrier to access for some groups, particularly low-income groups.

Scientists coming from different schools will approach the same policy problem, such as controlling hospital costs or assuring equitable distribution, quite differently depending on which of these sets of assumptions they use. Until recently there was little empirical data to support or challenge either set.

In assessments the challenge is to: clarify the issues; identify which aspects hinge on facts and which on values; decide what kinds of evidence will be determinative; grade the degree of certainty about points of applicable technical knowledge and the degree of consensus over points of social values; take note of how sensitive the factual analysis will be to issues meta to the fact-finding; then assess the overall situation. If wished, facts about social values may be adduced, benefits and costs of reducing uncertainty estimated, and social consequences appraised.

Assessors' values can be stated openly: if a committee states at the outset that it prefers "maximization of consumer choice," its recommendations on, say, food additive policy can be interpreted straightforwardly. Facts contingent on value preferences should be identified as such: "Dermatoses are the most serious occupational diseases affecting American workers" needs qualification. Nuclear optimist Edward Teller violated the distinction when he argued in his "I Was the Only Victim of Three Mile Island" advertisement in the *Wall Street Journal* (July 31, 1979) that "Nuclear power is the safest, cleanest way to generate large amounts of electrical power. This is not merely my opinion—it is a fact."

All of us need to get better at recognizing implicit value judgments, such as the weighting of concerns in this passive imperative on diesel emissions:[41]

Evaluating the potential pulmonary and systemic effects of exposure to diesel exhaust should be done with emphasis on anticipated morbidity [illness] rather than mortality.

41. NRC, Diesel Impacts Study Committee, 1982, p. 72.

And those in this study of reproductive hazards in the chemical workplace (Karrh et al., 1981, p. 399):

> Where there is potential for exposure to an embryo-fetotoxin for which an acceptable exposure level cannot be set due to inadequate data, women of reproductive potential should be excluded from the work area.

Both recommendations are straightforward, debatable fact–value hybrid propositions.

Constructing facts in an area of ignorance can itself raise value questions. Description of the unevenness of benefits and burdens to different social groups under the Clean Air Act, description of voter sociodemographics in an area being reviewed for election district rezoning, or confirmation of a specific cause of public health harm, can profoundly change the shape of a controversy. Where there was little focused contention before, options can be reappraised, and opportunities and liabilities levied. Only after overwhelming medical consensus, bolstered by political concern, established that pneumoconiosis (black lung disease) presents a major threat to coal miners did federal regulators attend to that hazard in addition to explosion and cave-in accidents (Key, Kerr, and Bundy, 1971). For many years until recently the cities around the Mediterranean Sea resisted letting the United Nations Environment Program collect and analyze their sewage and industrial waste outfall; the uncertainty masked culpability.

Factual Uncertainty and Indeterminacy

Factual uncertainty and indeterminacy raise other meta-analytic issues. Witness this testimonial on uncertainty at the conclusion of the National Academy of Sciences' heroic review of the widely used anesthetic, Halothane:[42]

> In the history of medicine, it is doubtful whether any drug was ever more extensively studied both before and after its introduction than Halothane. Yet, after Halothane had been given to patients ten million times, it was impossible to give firm, reliable answers to many basic questions about its effects. Two such questions were: How does the death rate after operations under Halothane anesthesia compare with death rates when other anesthetics are used? Does Halothane induce significantly more liver function impairment than other widely used anesthetics? The National Halothane Study attempted to answer these questions by using existing records. Although 856,000 opera-

---

42. NAS/NRC, Subcommittee on the National Halothane Study, 1969, p. 428.

tions were brought under scrutiny, the answers given are predictably and regrettably short of those desired. The limitations of knowledge on Halothane are certainly not peculiar to it. Limitations at least equally compelling apply to nearly any drug introduced in the past.

That situation has not changed much. Science is far from omniscient. Yet it is constantly being pressured for answers on problems like Halothane.

But uncertainty does not mean no knowledge at all. Some commentators make that assumption, thus overlooking one of the central powers of science: its ability to make, and then refine, estimates. That we do not understand precisely a chemical's proclivity to cause cancer does not mean that we understand nothing about it; we may well be able to estimate both upper-bound and average exposure–effect probabilities, and estimate how many people possibly could be exposed. That we don't understand much about *how* acupuncture works does not mean that we can't affirm *that* it works.

Uncertainty itself can be surveyed. Needing state-of-art estimation of probabilities in the causal chain from stratospheric flight to upper atmospheric pollution and ultimately to changes in temperature and ultraviolet radiation at the earth's surface, one study polled recognized experts systematically. Its questionnaire directed:[43]

> The decision analysis approach requires that you, as experts in your scientific fields, perform the uncomfortable task of assessing in a quantitative way the extent of the uncertainty you honestly feel, deep down inside. The initial goal is to derive a set of judgmental probability *distributions* for the key links in the causal chains. It is *not* the purpose of this study to use these probability estimates in a full-blown decision analysis for stratospheric flight, since that would require measurement of the benefits derived from such flight and consideration of value judgments regarding the relative weighting of risks and benefits. It *is* our purpose ultimately to provide the decision-makers in government with an assessment of the probabilities of possible consequences of stratospheric flight, in the form of honest judgmental probability. . . .

Questions included, for example, "Assess the concentration of nitrogen oxides in the absence of emissions by aircraft at 6–9 kilometers above the tropopause," and, "Given that a new engine will be developed over a normal development cycle for subsonic aircraft, what is the incremental cost (in 1974 dollars) to reduce the nitrogen oxide emission index from 18 down to 6 grams per kilogram?"

Granger Morgan and his colleagues at Carnegie–Mellon University have

43. NAS/NRC, Climatic Impact Committee, 1975.

systematized this approach and applied it to the estimation of transport and health effects of sulfurous pollutants from coal-fired power plants (l978a, l978b, 1982). Such exercises not only allow elicitation of a shadow consensus, but also they indicate how sensitive the estimates are to uncertainty (that is, how much the estimates would change if uncertainties surrounding various factors were reduced) and what decisional payoff would accrue if the uncertainty were reduced by further research and analysis.

Indeterminate questions—questions that not only have not been answered but that cannot be answered with confidence under present circumstances—are even more troublesome. No Earth-being can know the probability that spores or viruses hitchhiking on our outbound spacecraft could survive to contaminate other planets; yet the question cannot be avoided, and experts have been asked to estimate those hazards and recommend quarantine policy.[44] Because each generation of commercial pesticides is displaced every five or ten years by a next generation, and because human cancer is clinically manifested only ten to twenty years after exposure to carcinogens, we outsmart ourselves, in this aspect, by changing chemicals before human evidence becomes available for fully assessing carcinogenic potential. Because time cannot be rushed, nobody can know exactly how vitreous blocks of entombed high-level radioactive waste will age over hundreds of years.

"Trans-scientific" is what Alvin Weinberg has called issues that "can be asked of science and yet which cannot be answered by science" (1972b, 1977). Included would be experiments that are conceivable but forbiddingly tedious and costly, such as radiation assessments at such slight doses as to require testing on billions of mice (for statistical power), and those, such as measuring the strength of Hoover Dam, that are impossible because we would never want to stress the actual structure to its breaking point. The obligation for scientists, Weinberg argued, is "to make clear where science ends and trans-science begins." Fair enough. But no science is absolutely certain; in every field from astronomy to cardiology, prediction is daily work; the frontier into trans-science may be imprecisely or incorrectly marked; and with improvement of scientific tools, questions trans-scientific to one generation may become scientific to the next.

AN ARCHETYPAL ANALYTIC EXERCISE:
THE RASMUSSEN REACTOR SAFETY STUDY

During the mid-l970s a team chaired by Norman Rasmussen, an MIT professor of nuclear engineering, conducted for the Nuclear Regulatory Commission (NRC) a major probabilistic analysis of the hazards of commercial nuclear reactors. The foot-thick, $4,000,000 *Reactor Safety Study* (NRC's

44. NRC, Committee on Planetary Biology and Chemical Evolution, 1978.

document no. WASH-1400) and the debate over it will long stand as a richly illustrative and stimulating exercise (U.S. Nuclear Regulatory Commission, 1975). What I want to show in it is how strongly meta-analytic considerations and expert judgment influenced its outcome.

The whole Rasmussen affair was politically and technically contentious, and was flawed in several ways. Overall, however, the endeavor that began in 1972, unfolded through countless meetings, drafts, arguments, redrafts, criticism, meetings, hearings, and reports, and more hearings, and revised reports, has been extremely useful. The lessons from it, both about reactors and about assessments, deserve to be learned.

Most of the technical input to the Rasmussen study was not original with the committee but had been developed elsewhere. What WASH-1400 did was: critique a wide range of specialized data; delineate potential accident initiators, sequences, and physical consequences; estimate the likelihood of the various accident sequences; and then estimate the health consequences from those accidents. It also sought to compare reactor accident risks with everyday health risks, to place them in perspective.

The study's analytic framework and numerical estimates were so important that the report was subjected to fairly independent critique by many organizations, including the American Physical Society, the Union of Concerned Scientists, the Electric Power Research Institute, the Environmental Protection Agency, the Nuclear Energy Policy Study Group of the Ford Foundation, the staff and a special review committee of the Nuclear Regulatory Commission (Lewis committee), and a panel of the National Academy of Sciences.[45] All these reviewers have critiqued each other's reviews. The electric utilities, nuclear equipment manufacturers, and foreign governments have had their experts evaluate the assessments all along. The extensiveness to which these organizations argued about the report is the best evidence that such dry-as-dust studies matter.

The Rasmussen study was based on two related methods, event-tree analysis and fault-tree analysis. An event tree begins with an initiating event (such as a pipe break), ramifies through alternative succeeding events (such as whether backup systems work), takes account of potentiating and ameliorating events (such as whether physical barriers or electrical control systems operate properly or fail), and pursues all the branches out to their consequences (such as return to normal, or damage but safe takeover by auxiliary equipment, or rupture and release of radioactive material from containment). After a tree of possibilities is constructed, probabilities are assigned, as will be described shortly, to each of the branches. To calculate the chances of events' cascading to the branchtips the probabilities are

45. American Physical Society, 1975; Union of Concerned Scientists, 1977; Electric Power Research Institute, 1975; U.S. Environmental Protection Agency, 1976; Nuclear Energy Policy Study Group, 1977; U.S. Nuclear Regulatory Commission, 1978; and NAS, Committee on Literature Survey of Risks Associated with Nuclear Power, 1979.

multiplied out from the initiating event, through the intermediate events, to the final consequences. Inversely but in the same manner, a fault tree begins with some imagined final state of failure in the system (such as loss of electric control for part of the machinery) and ramifies backward up the causal chain to portray the various faults that could lead to such failure. Fault-tree and event-tree analyses complement each other in looking for trouble.

Such analyses not only help size up the risks, but they identify possibilities and priorities for reducing those risks. Remedy can be to upgrade weak-link components, build in redundancies, isolate some systems or connect others, modify operation, maintenance, or inspection procedures, or add damage-containment safeguards.

Complexion of WASH-1400

It is possible without wading through the entire WASH-1400 to recognize how deeply the study's design and assumptions—its meta-analytic dimensions—influenced the report's value-freighted outcome.

TECHNICAL DETAILS. There were, of course, myriad technical questions about pumps, pipes, valves, control rods, cables, and switches, about the behavior of materials at high temperatures and under irradiation and vibration, and about the interactions of automatic and manual controls.

COMPREHENSIVENESS OF ANTICIPATION. Could Rasmussen, or any other such study, have thought of all the things that can possibly go wrong? After all, reactors are staggeringly complicated, and assessment involves multiplying Murphy's law times Murphy's law times Murphy's law. . . . The crippling 1975 fire at the Browns Ferry plant in Alabama was touched off by an electrician's holding a candle up to a seal in a wall (in violation of standard procedure) to check for an air leak, which ignited the temporary sealant he had just patched the wall with, which smoldered through into the wall space, which then burned the plastic insulation off reactor control wiring that, in bad design, passed through the space alongside the two emergency backup cables, and led to such havoc as knocking out control systems and sending acrid smoke and fire-extinguisher carbon dioxide into the control room; much damage was done, and the accident was extremely threatening. A problem highlighted by the Three Mile Island accident is the great difficulty in anticipating operator error.

The Rasmussen team polled experts for months, trying to anticipate all such possibilities, and it enlarged the list of things that can go wrong. But there was not then, nor can there ever be, a way of proving that absolutely all possibilities have been taken into account.

COMMON-MODE FAILURES. Related to this is the question of joint or "common-mode" failure—nearly simultaneous, additive failure—of systems, as might be experienced if a hurricane severed electricity supply lines while several reactor systems were malfunctioning, or if an earthquake disabled a number of components all at one time. Uncertainty is compounded by the possibility of linked failures. The Rasmussen report didn't deal with these sequences very well, but attention has been devoted to them since then.

GUESSTIMATION OF PROBABILITIES. Some reactor components had been under enough observation, in nuclear or other industrial applications, that reliability statistics (such as those describing the likelihood of a switch's sticking) were available on them. But because of the newness of the reactor industry, many systems had accumulated only short operating experience. On these, experts were asked to estimate failure rates "subjectively," based on their best engineering judgment. Naturally this invited criticism. On this point, within the framework of the study and its admitted intrinsic limitations it is probably fair to say, as the NRC's Lewis panel did in reviewing WASH-1400:[46]

> *Reactor Safety Study* had to use subjective probabilities in many places. Without these, *Reactor Safety Study* could draw no quantitative conclusions regarding failure probabilities at all. . . .
>
> It is our view that use of subjective probabilities is necessary and appropriate, and provides a reasonable input to the *Reactor Safety Study* probability calculations. But their use must be clearly identified, and their limits of validity must be defined.
>
> It is true that a subjective probability is someone's opinion. But . . . some people's opinions can be very accurate, even in a quantitative sense. So the question really being raised is: how valid are the opinions of those whose subjective judgments were used in *Reactor Safety Study*? It must be kept in mind that the subjective probabilities used are for some of the individual steps in a sequence of events. No one provided a subjective probability for the overall probability of core melt. For many of the steps in which a subjective probability was used it was the output of experienced engineering judgment on the part of people familiar with events of that type. This, of course, does not guarantee the accuracy of the probabilities so generated, but if properly chosen makes them the best available. The only situation in which one might be concerned about a subjective probability leading to really major errors is where there is no experience at all. An example of this in *Reactor Safety Study* is nuclear pressure vessel failure. Such cases should be clearly identified, and results sensitively dependent on them treated with extra caution.

46. U.S. Nuclear Regulatory Commission, 1978, p. 9. Apostolakis, 1981, discusses such so-called "Bayesian" refinement of estimates.

HANDLING OF UNCERTAIN STATISTICS. Much debate has centered around WASH-1400's depiction of the frequencies of component failure and expressions of how-far-from-failure systems are likely to be (a notion familiar to any owner of a middle-aged automobile). Again, because of the briefness of the industry's history, the recordbooks yielded only modest information on variability of component failure rates. The Rasmussen committee recognized that it made many unsubstantiable assumptions about the chance distribution of breakdowns. It repeatedly discussed this problem. Arguing, however, that most systems behave in a manner mathematically described as "log-normal," for most calculations and fitting of measured data to ideal graphs WASH-1400 assumed that such distributions were not bad approximations of reality. Most critics have agreed with the Lewis review, which said, "Within the errors listed in *Reactor Safety Study*, we accept the log-normal as an acceptable summary of most data."[47]

BIAS IN THE FAULT TREES. The Rasmussen study unavoidably had to make thousands of working assumptions as it went along, and it usually chose to "err" on the conservative side, such as assuming that when control rods fail to drop into place they fail to drop at all, and that partial melting of the reactor core always leads to complete core-melt. Multiplying out these conservative subprobabilities can be said to have induced a conservative, or worst-case, bias in the overall assessment.

ESTIMATION OF PUBLIC HEALTH CONSEQUENCES. After assessing the chances of all the various accidents, the Rasmussen report went on to estimate their implications for public health. It had to make assumptions about how many people would be in the vicinity of the plant, how fast they would evacuate, and which way and how hard the wind would be blowing. All of these will be different from reactor to reactor. Besides, in modern times Americans have never evacuated a major city, so there is no experiential base.

Rasmussen estimated typical immediate fatalities and economic losses to be expected from reactor accidents. Its estimates of delayed illnesses, genetic damage, and delayed fatalities, as from radiation-induced cancer, have been criticized heavily. Such appraisals of course depend on hypothetical models and estimates.[48]

JUDGMENT OF MOST THREATENING ACCIDENT POSSIBILITIES. "Most threatening" refers, of course, both to estimated risks and to emotional apprehensions over risks. WASH-1400 argued that in previous assessments too

---

47. U.S. Nuclear Regulatory Commission, 1978, p. 9. For epistemological dissent, see the appendix to Levi, 1980.

48. Critique and alternative estimates are presented in NAS, Committee on Literature Survey of Risks Associated with Nuclear Power, 1979.

much emphasis had been allotted to extremely unlikely but catastrophic accidents leading to reactor core destruction, and not enough attention to somewhat more likely, dangerous though less catastrophic, accidents, such as those resulting from stuck-open valves or from breaks in small pipes (Three Mile Island exemplified the latter). Current design analyses are shifting this emphasis.

In Rasmussen and elsewhere the argument has been made that just as the public feels greater anguish over sudden loss of many lives (as in an airliner crash) than it does over diffuse loss of the same number of lives (as in the automobile wrecks that occur every day), it would be greatly traumatized by an enormous reactor accident, even though the event were extremely rare. For that reason nuclear power plants have been designed to be disproportionately protective against catastrophic accidents (Griesmeyer, Simpson, and Okrent, 1979).

CONCEPT OF "ACCIDENT." In the opening hours of the Nuclear Regulatory Commission's emergency conference when the Three Mile Island (TMI) reactor took its "excursion from normality" in March 1979, Commissioner Richard Kennedy asked, "Is this an accident? What is an accident?"[49] Managers of large-scale technological enterprises grow accustomed to a notion of routine "transients," or "design-basis, expected accidents"—that is, accidents that sooner or later are almost bound to happen. The line between expected and unexpected variation, and between normal variation and accident, is fuzzy.

Looking back on TMI, early observers tended to hold one or the other extreme view: either that TMI had gotten completely out of control and was the worst reactor accident in history and a disaster; or conversely that since nobody, or perhaps in the long run no more than a few workers, had been hurt, the system had convincingly proved its safety (Edward Teller proclaimed that TMI demonstrated that "nuclear reactors are even safer than we thought" [1979]). Surely the "truth" is somewhere in-between.

The analytic question is, How near a miss was the TMI accident? The Kemeny Commission staff and consultants examined this in detail. A Panel on Alternative Event Sequences (Stratton panel) carefully went through many "What if?" cases and concluded that "No single additional operator action or equipment failure that is tied to the actual sequence of events at TMI would have led unequivocally to large scale fuel melting throughout the core or significantly larger release of fission products to the environment."[50]

The TMI episode was held up to the Rasmussen report in a number of ways, by many different groups, asking: To what extent and in what sense

49. Transcript of Nuclear Regulatory Commission/IRC telephone logs, no. 01-01271-CH7-25-EH-5, March 28, 1979.

50. U.S. President's Commission on the Accident at Three Mile Island, 1979b, p. 7.

did the *Reactor Safety Study* predict TMI? In the Kemeny review, another technical staff analysis (Burns panel) concluded that although some aspects of the accident sequence had not been anticipated, "WASH-1400 is remarkably defined and accurate in its description of events corresponding to those which occurred at Three Mile Island," and that its "upper bound probabilities predicted an 80% chance of having had an accident this soon" in the industry's history. "We should have anticipated TMI-2 and been prepared to deal with it."[51]

IMPLICATIONS OF GENERIC OVERVIEW FOR INDIVIDUAL REACTORS. The Rasmussen assessment was modeled on several "typical" American reactors. But around the country actual designs and sites differ greatly, and individual reactors change over time through different maintenance and operation. So the implications of WASH-1400 for any real plant can always be questioned.

OVERALL SCOPE OF ASSESSMENT. Even though the issues were not included in its charter, the Rasmussen study has been faulted for not examining reactor risks that might arise from sabotage, theft of nuclear material, or terrorist abuse. The committee acknowledged these omissions, but said that such issues were even less amenable to quantitative analysis than those it was struggling with already. Since then, terrorist hazards have been reviewed in several published studies, and they also have been scrutinized in unpublished industry and classified government studies.

David Okrent, a UCLA professor of engineering and applied sciences, has said, "The study of the way fires contribute to risk, a very difficult subject to analyze, is inadequate in the WASH-1400 report [and it] grossly underestimated the potential of earthquakes to risk from light-water reactors. And it may have missed out on floods."[52]

The Rasmussen committee's charter restricted it to reactor safety and did not include risks from the rest of the nuclear fuel cycle, such as uranium mining, fuel processing, waste disposal, worker exposure, or decommissioning of plants at the end of their operating lives. The report did not compare nuclear to other sources of electric power, although of course it has been and will continue to be used extensively in such comparisions.

A last caveat about the Rasmussen report. Probably in its own terms the *Reactor Safety Study*, as modified, now is reasonably accurate. In quite other terms, however, the study just may not have asked all the socially crucial questions: It is not impossible that the social repercussions of all the accidents summed for the long term by WASH-1400's tables could be outweighed by a single reactor takeover by terrorists, or that an accident kill-

---

51. U.S. President's Commission on the Accident at Three Mile Island, 1979c, pp. 5 and 17.

52. Okrent, 1981, p. 319. Okrent's book provides important perspective on the Rasmussen study and its context.

ing "only" several thousand people and contaminating a metropolitan area could shut down the entire nuclear electric industry in America.

Usefulness of the Exercise

Limitations notwithstanding, major assessments like the *Reactor Safety Study* can serve powerfully as focusing mechanisms, help identify weak links in the protective chain, and probe uncertainties. The Rasmussen study has been exhaustively (and exhaustingly) reviewed, revised, qualified, reinterpreted, and evaluated against experience. Its legacy continues in countless analyses and reactor systems. What remains important is not the document but the entire Rasmussen exercise.[53]

53. General commentaries include Lewis, 1980; Okrent, 1981; Okrent and Moeller, 1981; Moss and Sills, 1981; Norman McCormick, 1981; Ford, 1982, pp. 133–173; and U.S. President's Commission on the Accident at Three Mile Island, 1979. Mynatt, 1982, describes research trends since TMI. Sills, Wolf, and Shelanski, 1982, is a symposium of social-science commentaries on TMI. U.S. Nuclear Regulatory Commission, 1984, broadly reviews probabilistic assessment of reactor risks. *Risk Analysis* 4, pp. 247–335, 1984, is a special issue on nuclear probabilistic risk analysis.

Eight

Taking Account of Social Values and Ethics

"To shape the world nearer to the heart's desire," remarked Walter Lippmann, "requires a knowledge of the heart's desire and of the world" (1914, p. 269). Yes. But how to know *society's* heart's desires?

The preceding chapters have described a great many world-shaping activities in which values need to be known and applied: in guiding professional judgments and practices; in setting R&D priorities; in regulating experimentation; in formulating assessments; in shaping and applying technologies.

Much more frequently than they may expect to, technical leaders and advisory committees find themselves drifting into—and then becoming immersed in—heated discussions of personal rights, or "the natural," or obligations to future generations, or John Rawls's theory of justice, or the ethical dimensions of cost–benefit analysis. Often these considerations turn out not to be peripheral, but central. At issue may be the substance of decisions, or institutional or professional roles, or social procedure.

SOURCES OF ETHICAL GUIDANCE

Most people find it very difficult to say, in specificity and detail, what their own values are. In the United States and other industrialized Western countries no unified morality prevails. Judeo–Christian precepts have taken on a diversity of secular interpretations, and have been adapted only awkwardly to recent sociotechnical progress. No one traditional religious system, or ethical system derived from modern European philosophies, dominates. Jeffersonian and other conceptions of the good society pervade law and policy, of course, but they endure in the public mind mainly as rules and pop-mythical slogans.

The organized religions have not helped much. "It is surprising," Abraham Edel has observed, "given the traditional strength of religion in Western civilization, what a small role religious concepts have played for the past two centuries within the inner texture of ethical theorizing" (1970, p. 25). Of his theologically trained colleagues who write on ethics, James Gustafson has complained that "the relation of their moral discourse to any specific theological principles, or even to a definable religious outlook, is opaque" (1978, p. 386).

Many new issues are stressful to traditional doctrine. Jewish leaders struggling to derive guidance on transsexual surgery have had to reach all the way back to Leviticus's prohibition, "That which is mauled or crushed or torn or cut you shall not offer unto the Lord" (22:24), and Deuteronomy's "A woman shall not wear that which pertains to a man, nor shall a man put on a woman's garment" (22:5). Rabbinical scholars trying to relate psychotherapy to traditional values must reconcile it with notions of healing—"And you shall return him to himself" (Sanhedrin 73a)—and *hokhahah*, or ethical rebuke, and *viduy*, confession. The Talmud still anchors the bibliography of these scholars' essays.[1]

Roman Catholic leaders, recognizing that at death "the 'call of God' increasingly comes to us through human voices and human hands," are troubled over what "meaningful, fully human life" means in artificial-life-support cases like Karen Quinlan's, and over whether Catholic interpretations are consistent with the civil courts' criterion of "cognitive, sapient" person. Catholic commentators are asking whether in vitro fertilization (test-tube conception) will erode emotional bonding and family life, "whether it amounts to treating the child as a possession of the parents rather than as a developing person with his or her own destiny," and whether it will deny "the right of a child to its own parents."[2] Except on such intensely felt and widely publicized issues as abortion, parishioners seem not to be much engaged.

Protestant churches and writers have not explored or developed guidance in depth on many of the new bioethical issues.[3] From time to time they have, however, voiced concern over broad societal problems. The 1983 General Assembly of the United Presbyterian Church endorsed a statement that "The church is called to seek justice in the delivery of health care in our increasingly complex society. At the same time, injustices in the

1. Rosner and Bleich, 1979, compiles essays on Jewish bioethics. See also Franck, 1983, and Spero, 1983.

2. McCormick, 1981, pp. 352–361 and elsewhere; this book is a good source on Catholic views, as is McCormick, 1984. Ashley and O'Rourke, 1982, broadly explores Catholic bioethics.

3. A recent exception is Lebacqz, 1983.

current system which are being exacerbated by deep cuts in federal funding must be challenged." It went on to urge that "access to a basic, minimal level of health care be available to all persons" and that the Congress "pass legislation which would enable a cost effective, yet fair, health care system to be established" (*Church & Society* 75, no. 3, p. 4, 1984). Whether such resolutions ever result in political effect is not evident.

Recently churches have referred some important questions to the secular polity. In 1980 the National Council of Churches, the Synagogue Council of America, and the United States Catholic Conference together petitioned President Reagan to task his bioethics commission with exploring the ethical issues surrounding genetic engineering; this led to the commission's report, *Splicing Life*. A central section of the report responds to "concerns about 'playing God.'" Although the report helped define the arena, it provided little specific guidance on upcoming ecological intervention and human genetic-modification proposals.[4]

Conflict between state and religious practices often forces painful confrontation. Controversy recurs over whether Amish children should have to submit to polio vaccinations and other public health protections (for their neighbors' benefit as well as their own), and over whether Jehovah's Witnesses and other fundamentalists who deny their children lifesaving medical care, such as insulin for diabetes, should be immune from felony charges. Right now debate is raging over whether the expenses of Christian Science spiritual healing should be reimbursed, as they are, from Medicare and other public insurance funds.[5]

Over the years, national commissions and similar groups have served constructively to focus debate and express the tenor of public opinion. Since their mandate is not to make policy but to draft an essay to the polity and make recommendations, they can gather diverse opinions and engage in lively, protracted, nonlegalistic discussions. The work of the National Commission on Protection of Human Subjects mentioned in Chapter Six was a good example.

Those national consensus groups tend to beg off from addressing the most fundamental questions. In the summary of its elaborate study, the more recent President's bioethics commission excused itself this way:[6]

> In the Commission's analysis, three basic principles predominated: that the well-being of people be promoted, that people's value preferences and choices be respected, and that people be treated equita-

4. U.S. President's Commission for the Study of Ethical Problems in Medicine and Biomedical and Behavioral Research, November 1982.

5. Swan, 1983, and Nathan Talbot, 1983, raised the issues, and letters of response appeared in the *New England Journal of Medicine* 310, pp. 1257–1260, 1984.

6. U.S. President's Commission for the Study of Ethical Problems in Medicine and Biomedical and Behavioral Research, March 1983d, p. 66.

bly. The Commission has not undertaken to rank these three principles among themselves nor to set them above other values—such as efficiency and honesty—that were less prominent in the particular studies it conducted. Although it has tried to apply these principles consistently in its various reports, the Commission has made no attempt to develop a comprehensive theory of bioethics: its assignment from Congress was not to develop theories but, more practically, to consider the implications of particular practices and developments in the life sciences.

Pragmatics, not theories. In an article in the *New England Journal of Medicine* the chairman of that Commission and an associate explained to doctors in lawyerly, common-denominator-seeking language (Abram and Wolf, 1984, p. 630):

> Membership diversity in expertise and politics rendered agreement unlikely not only on a particular case but also on an abstract ethical theory. This encouraged abandonment of any attempt to reach agreement on the level of ethical theory, since consensus was a desideratum. Moreover, the commission could not be dominated by ethical theorists, because the statute's membership provisions permitted no more than five professional ethicists among the 11 commissioners and required none.

Dismayingly, although one can't help appreciating the panel's difficulties, the chairman's tone is a bit too much "Moses, take those abstract-theoretical tablets elsewhere, and Hippocrates, keep your lectures to yourself; for bioethics need not be based on ethics." And is it being implied that only professional ethicists think about compassion, charity, justice, dignity, and freedom? Actually, the Commission's reports assembled a good bit of ethical thinking, even if in appendices, and will serve as prologues to further deliberation. They need to be critiqued and bolstered by contributions from civic, religious, and professional groups, and from scholars and journalists.

Obviously the civil and criminal laws reflect and consolidate society's ethical consensus, but, again, technical innovations may leap far beyond legal precedent. Searching for what laws might govern the sale of a living person's kidney or other organ, George Annas had to conclude that the most applicable statutes would be those relating to mayhem, assault, battery, maiming, and medical negligence (1984). Many such challenges are arising. The laws hardly are evolving apace with the technologies.

Mark, nonetheless: few, if any, civilizations in the past have done better; none has aspired to the extraordinarily diverse and egalitarian ambitions and freedoms that contemporary societies do; and few have managed to foster values-unity without having to enforce it centrocratically and violently. Nostalgia offers no solace. The civic challenge is enormous.

Tendentiously and ever so slowly we do, I believe, make ethical progress. We do continue to subject public decisions to ethical scrutiny. We do adapt societal and personal goals and preferences to changing physical and social exigencies—not always, and not quickly enough, but fairly well over the long term. We do continue, with at least moderate success, to impose our values on physical and social change. Progress occurs in the slow grinding of governments, in the slow working of corporations, and in the slow pondering of the courts, all influenced by countless personal and organizational forces.

The pervasive ethical questions remain, as ever: How should choices be made when options cannot all be pursued or when they conflict? How should collective societal goals be pursued with least erosion of the rights and goods of affected individuals? And how should specific guidance on particular concrete actions be derived from abstract high precepts?

"VALUES" AND "VALUE," AGAIN

At the opening of this book I pointed out that values range from broad, idealized virtues and goals, to specific, attainable objectives, and that *ends*, *means*, and *modes of conduct* are accorded value.

The "value [of something]" I took to be ascribed worth, reflected in social transactions. I said that I found actual manifestations of valuation— in political actions, budgets, laws, treaties, regulatory policies, military strategies; in court, corporation, and labor union decisions; in consumer purchasing behavior; in medical preferences; and in wage differentials and insurance schemes—much more telling than "opinions."

"Values" are abstract ideals. Despite common conversational reference to "traditional Midwesterners' [or Laborites' or Aleuts'] values," it is hard to define the concept. Social scientists have struggled to clarify it. Anthropologist Clyde Kluckhohn and his colleagues said that a value is "a conception, explicit or implicit, distinctive of an individual or characteristic of a group, of the desirable which influences the selection from available modes, means, and ends of action" (1967). In the view of psychologist Milton Rokeach, "To say that a person has a value is to say that he has an enduring prescriptive or proscriptive belief that a specific mode of behavior or end-state of existence is preferred to an opposite mode of behavior or end-state" (1973, p. 25). Sociologist Dennis Foss defined values to mean "beliefs about classes of objects, situations, actions, and wholes composed of them in regards to the extent that they are good, right, obligatory, or ought to be" (1977, p. 112).

Countless classification schemes have been proposed. Some of these distinguish ultimate, ideal values from mundane, operational values. Some distinguish essential-needs values from luxury-desires values. Some distinguish state-of-attainment (ends) values from dynamics-of-pursuit (instrumental) values. Some distinguish self-oriented values from group-oriented values. Some distinguish positive (desires) values from negative (aversions) values. And some categorize values by content or source: aesthetic, economic, religious, political.

(Jeremy Bentham set the precedent for all of this in 1815 with his elaborate Table of the Springs of Action, "Shewing the several Species of Pleasures and Pains, of which Man's Nature is susceptible: together with the several Species of *Interests, Desires,* and *Motives,* respectively corresponding to them. . . ." In his Table Bentham enumerated pleasures/pains associated with such Springs as the palate, sexual appetite, pecuniary interest, power, curiosity, amity, and self-preservation. He proposed that the values so classified become the terms for a "felicific calculus," which somehow would sum them, taking into account their intensity, duration, certainty, immediacy, and other properties.[7])

Whatever the definitional specifics, *to value* something—to hold interest in it (or in disparaging or avoiding it), to approve of it, desire it, be disposed toward it, prefer it to alternatives—is taken to be derived from, and reflect, abstract ideal *values.* Values are, as a minimum, personal mental constructs (such as desire for freedom), and they may, if assented to mutually by the members of a society, become adopted as societal constructs (ethically sanctioned, politically championed, legally guaranteed freedoms). The values/valuing relationship is a chicken/egg relationship; each gives rise to the other. Values serve as criteria for valuing: to help a child valuate properly [in some sense] in future instances, one instills general cultural values in him. Inversely, successions of valuative judgments can be taken to indicate values: if one wants to know a group's values, as one might in trying to resolve a controversy, one may inquire into how those people have acted and spoken on similar issues in the recent past, and how they value various factors in the present situation (recognizing that how they do value is not necessarily how they will or should value those things).

Several mid-century American presidents pressed for systematic reinterpretation of national goals, in the hope that this would encapsulate core values. Eisenhower appointed a national commission to prepare the report *Goals for Americans*; it made for stimulating highschool reading (I was affected by its vision), but probably it affected national policy not a whit (U.S. President's Commission on National Goals, 1960). Kennedy and Johnson defined their own aims for the nation. Nixon appointed a

---

7. Bentham outlined the proposal initially in his *Principles of Morals and Legislation*. The Table was published as Bentham, [1815]. Later Bentham would write, plaintively: "'Tis in vain to talk of adding quantities which after the addition will continue distinct as they were before, one man's happiness will never be another man's happiness . . . [but it is] a postulatum without the allowance of which all political reasoning is at a stand. . . ." (quoted from a manuscript, in Halévy, 1901, vol. 3, p. 481, note 55).

National Goals Research Staff in the White House, which prepared a report, *Toward Balanced Growth: Quantity with Quality*; but that document examined not so much goals as instrumentalities for achieving such traditional goals as "equal job opportunity" and "environmental protection amid economic growth" (U.S. National Goals Research Staff, 1970). No comparable study has been undertaken since.[8]

Social scientists repeatedly have attempted to survey people's values. Let us traverse a brief blizzard of inverted commas—the value descriptors are heavily definition-qualified—to take notice of the analytic problems and the diversity of classification themes devised.

Florence Kluckhohn and Fred Strodtbeck analyzed how the inhabitants of five neighboring Southwestern ("Rimrock") communities—a Texan homestead community, a Mormon village, a Spanish-American village, a Navaho band, and a Zuñi pueblo—were oriented to "human nature," "man–nature," "time," "activity," and "relational" values (Florence Kluckhohn and Strodtbeck, 1961).

Nicholas Rescher proposed a framework for analyzing value change, and suggested classifying values on such scales as "personal material welfare," "skill," "domestic virtues," "justice," "patriotism," and "environmental beauty." But Rescher never conducted a full-scale survey (1969).

Robin Williams, defining values as "conceptions of desirable states of affairs that are utilized in selective conduct as criteria for preference or choice or as justification for proposed or actual behavior," surveyed Americans' values and found them to emphasize "achievement and success," "moral orientation," "efficiency and practicality," "external conformity," "secular rationality," and "democracy" (1970).

Milton Rokeach elaborately surveyed Americans' rankings of such "instrumental values" as "ambitious," "honest," "loving," and "self-controlled," and such "terminal values" as "an exciting life," "a world of beauty," "national security," and "true friendship" (1973).

Ronald Inglehart analyzed European Community surveys and concluded that between the 1960s and the 1970s European values underwent "a shift from overwhelming emphasis on material consumption and security toward greater concern with the quality of life" (1977).

And so on. Many such surveys have been published. Still, it is very hard to know how to apply such generalized findings to real-world decisions, except to recognize that they, along with other newsmagazine-level generalizations, may affect the tenor of personal feeling and public debate.[9]

One attempt at applying surveyed values to public decisions was the 1970s Ohio River Basin Energy Study, an "assessment of the potential environmental, social, and economic impacts for the year 2000" expected

8. A good review of these projects is U.S. Congress, Congressional Research Service, 1971.

9. Clyde Kluckhohn, 1958, summarizes mid-century surveys of American values by such scholars as Ralph Barton Perry, Arthur Schlesinger, Sr., and Robert Lynd.

from construction of coal-fired power plants in that six-state region. The study reviewed published analyses of American values, assembled them into a composite profile ("economic benefit," "nationalism," and so on), and estimated how these values would be reflected in different energy-growth scenarios (such as "conventional technology/lax controls/high growth") in the basin (Potter and Norville, 1981).

Perhaps generalizations are the best we can expect; certainly that's what we tend to get. As I mentioned above, in 1982 the President's bioethics commission summarized the core American health-decision values as "informed consent," "serving the patient's well-being," and "respecting self-determination."[10] In that same year American environmental values were summarized by Donald McAllister as "preserve nature," "conserve resources," and "control pollution" (1980, p. 58). (But what is meant by "nature"? And since a way to conserve New England forest resources and reduce smoke pollution is to replace wood-fired heating with nuclear-generated electric heating, does this make nuclear power highly valued environmentally?)

APPRAISAL OF PEOPLE'S VALUES

Systematic assessments, I have argued, can help organize and inform decisionmaking. These assessments in turn depend on appraisal of people's values: how much they will pay for a dental benefit, how much automotive speed they will forgo in order to safeguard children in a school zone, how hard they will press for fuller employment. For decisions, how can value be assigned to various options and consequences?

Elicitation of Health and Environmental Preferences

In some cases, people affected by a decision simply can be asked their preferences. Barbara McNeil and her colleagues in Boston medical centers have interviewed both healthy and ill people about cancer-treatment options. Subjects told to imagine themselves with bronchogenic carcinoma and having to choose between surgical extirpation, which carries a small chance of immediate death during surgery, and radiation therapy, which on average is less life-prolonging but carries little risk of immediate death, generally preferred the modest but surer survival promised by the radiation option (McNeil, Weichselbaum, and Pauker, 1978). In another study, some patients afflicted with cancer of the larynx preferred radiation therapy, which preserves speech but shortens survival, over surgery, which improves survival but at the loss of speech. "These results suggest that

10. U.S. President's Commission for the Study of Ethical Problems in Medicine and Biomedical and Behavioral Research, October 1982.

treatment choices should be made on the basis of patients' attitudes toward the quality as well as the quantity of survival," the investigators concluded.[11] Patient-choice factors in other situations could include pain, hospital confinement, disfigurement, or loss of mental acuity. Preferences reflect comparative valuation.

A variety of studies of health values are being performed now. Some progress is being made, for example, in surveying workers' attitudes about skin rash as compared to loss of a finger, loss of a finger compared to loss of an eye, and so on. The work is very difficult. It suffers from all the usual limitations of polling, and some surveyors report that quite a few respondents break off the interview because the questions are emotionally disturbing. Too, it is not evident that patients actually facing medical problems will choose in the way these health-status indices predict.[12] Although surely individuals can, and should, be asked their medical preferences, it is very hard to aggregate these into "society's" preferences, except coarsely. Variance among people is likely to remain high.

Local environmental preferences may be easier to elicit. Residents' reactions to different aircraft noises have been surveyed, as have people's opinions of different forest logging patterns, National Park landscapes, and chemical pollutant odors. Voters have expressed themselves in referenda on chemical and radioactive-waste disposal, oil tanker unloading facilities, home solar energy zoning, and glass-bottle recycling, and have passed resolutions of concern about nuclear power and acid rain. A major problem is that locally expressed preferences may be inconsonant with national or other collectively formulated objectives.[13]

Appraisal of Life-Preserving Investments

The appraisal of societal life-preserving actions is a vexing problem for medical and public health programs and for most technological programs, from building codes to tsunami warning systems to beer-bottle quality control regimens.

Cost–benefit and other decision approaches are *not* "pricing life" as though bodies were being auctioned. Occasionally journalists lugubriously molest such analysis, as Michael Brown did in an article entitled "The Price of Life" in *New York* magazine (March 10, 1980). In that article an elderly woman's photograph was captioned "Senior citizen, Brighton Beach: $60,000," a "Scarsdale stockbroker" was tagged "$1 million," and a large header intoned, "The state has quietly calculated, in dollars and cents,

---

11. McNeil, Weichselbaum, and Pauker, 1981. See also McNeil et al., 1982.

12. Read et al., 1984, compares different methods for assessing health-outcome preferences.

13. Andrews and Waits, 1980, reviews environmental values research. Sinden and Worrell, 1979, assembles many environmental decision examples.

whether it's worth spending tax money to save your life." This distorts the point. The point is that since we constantly expend money and effort on reducing risks and enhancing quality of life, evaluation can help us recognize what we actually are buying and decide how we might want to change those investments, most of which begin as responses to emergencies, or as routine bureaucratic actions, and subsequently just grow, unexamined. Human life may be beyond price, but food inspection programs, cardiac emergency facilities, and bridge maintenance regimes aren't.

Usually the sort of analysis Brown was disparaging involves reasoning that, for example: if everyone in a group pays $250 to avoid a hazard that otherwise would take the life of about one person in a thousand, that amounts on statistical average to $250,000 spent per life saved. Such analysis might reveal that even as the group willingly pays for that protection, it is overlooking another possible investment that would cost only $80,000 per life saved. The situations exist, whether analyzed or not.

Facing mortal facts can be unsettling—which is, I suppose, why analyses are often misunderstood and misconstrued. All such analyses are difficult and imprecise. They should be subjected to searching critique. But to ridicule the very idea of trying to compare life-preserving investments is intellectually and ethically Luddite.

In 1980 James Vaupel published an accounting of prospects for saving lives in the United States that exemplified how empirical analysis can appraise "the way things are" and raise issues of justice (1980, p. 101):

> [The statistics reviewed] indicate that the aggregate social losses due to death are largely attributable to early death and that the losses due to early death are immense, that the early dead suffer an egregious inequality in life-chances compared with those who die in old age, and that non-whites, the poor, and males suffer disproportionally from early death. Furthermore, statistics on the leading causes of death and statistics comparing non-whites and whites, males and females, current mortality with mortality earlier in this country, and the United States with Sweden and other countries suggest that early deaths could be significantly decreased.

Vaupel's study pointed both to inequities and to practical opportunities for prevention of early deaths.

Two main approaches have been taken in appraising life-preserving investments. Neither is very satisfying philosophically. Neither claims to weigh either personal or societal losses fully. But in context with a variety of comparative assessments, they can help estimate the relative payoff of contemplated actions.

One is the *human capital* approach. As far back as the mid-1600s William Petty simplistically calculated monetary return to the English economy that would accrue from "advancement in the medical art" or "lesening ye plagues of London," costed to the price of slaves or to laborers' earnings.[14] Modern human-capital analyses usually sum up net contributions to the social economy, including future earnings saved and medical and other costs reduced, that are expected from some health enhancement. Clearly such an approach yields only a very minimal estimate of burden to society, and is unable to capture the personal costs of pain, lost opportunity, grief, and other important intangibles.

The other basic approach, *implicit life valuation* or *willingness-to-pay analysis*, infers values from marketplace, insurance, court liability, and other actions that purchase protection or compensation for life- or health-deprivations.

Polling or mock auctions are sometimes used to gauge people's willingness to purchase protection. But although willingness revealed in these ways incorporates some sense of medical and other costs avoided and suffering averted, people's intuitive estimates of cost may not be accurate. There is no guarantee that people's responses to surveys are a reliable index to what they will actually pay in the market, or what, after incurring harm, they might wish they had paid.

Wage premiums for especially hazardous work can be taken as risk valuations, as can black lung and other compensation schemes.[15] But since wage markets are not necessarily either fully informed or freely bargainable, it is difficult to draw inferences from them. It is questionable whether it is valid to project any such derived valuations into other workplace or nonoccupational situations. Court liability awards can be analyzed, but often courts don't reveal the basis on which fines were set, and they may add heavy punitive and "demonstration" amounts into the awards. Purchasing behavior (such as willingness to buy household smoke detectors) and insurance investments are also revealing. But these expenditures may be based on inaccurate risk estimates on the part of the purchasers, and may be influenced by a variety of nongeneralizable factors (such as inordinate fear of fire) or catastrophe aversiveness (such as parents' apprehensiveness about leaving children destitute if they should die early). Again, it is hard to project valuations from one activity into another. Some analysts now are trying to hybridize revealed-preference with human-capital valuations.[16]

14. Fein, 1971, succinctly traces the history of human-capital analysis in the health field.

15. Barth and Hunt, 1980, is a good review of workers' compensation in the United States, seven European countries, and Ontario.

16. Landefeld and Seskin, 1982, provides a succinct review of both approaches, and attempts a hybridization. Jones-Lee, 1976, Mooney, 1977, and Menzel, 1983, provide overviews. Rhoads, 1980, compiles classic papers on both approaches. Institute of Medicine, Committee for a Planning Study for an Ongoing Study of Costs of Environment-Related Health Effects, 1981, discusses both approaches.

Increasingly, the payoff from public programs in forestalling premature death is coming under critical appraisal. Coalminer self-rescue devices, railroad crossing designs, kidney dialysis programs, coke-oven emission controls, measles vaccination campaigns, and many other programs have been analyzed. Nonfatal-injury preventives also have been studied. Most appraisals have been imprecise. Many ignore discrepancies among the benefits the sources of hazard provide. But taken together they are beginning to reveal wide differences in efficiencies among programs.[17]

As a result of vigorous discussions of these matters in government, insurance industry, and other circles, life-preserving investments are tending to be brought toward consistency with each other. No doubt this will continue. Comparative analysis is strongly affecting the practice of medicine, as it scrutinizes questionably efficacious and extremely expensive medical procedures and professional habits. And comparative analysis is causing engineers to rethink many traditional design practices, such as the imposition of large and expensive "safety factors" having nearly arbitrary origins with little analytical justification, and to reemphasize inspection and maintenance of constructions.

PHILOSOPHICAL UNDERPINNINGS OF SOCIAL ETHICS

Who may have what?, *Who may do what to whom?*, *Who should do what for whom?*, and, in each case, *Why?* are questions that have to be readdressed by every people in every generation. I can't possibly settle them here. Many of the ethical questions surrounding science and technology are subsets of perennial issues of morality, justice, and the good life. But some others are completely new, or are troubling variants of older issues, or deeply involve technical people; it is these that I would like to draw attention to. I will begin by précising some fundamentals of rights-based arguments, utilities-based arguments, and liberal egalitarian arguments.

"Rights" are a fundamental, but in many ways insubstantial, basis for moral and ethical principles. The great burgeoning of rights claims came, of course, in the revolutionary declarations of the eighteenth century, such as the United States Bill of Rights. America was served well in its early days by the premise that rights, "self-evident" and "inalienable," exist somehow intrinsically in the moral order, bestowed in Nature by Divinity. The secularization and modernization of society since then has steadily transmuted those notions, however, leaving us with a legacy of customarily granted rights. Those rights are claims asserted by individuals either

17. Graham and Vaupel, 1981, compares cost-per-life-saved by different federal programs.

against other members of society or against society as a whole. Some are rights *from* various interferences, and some are rights *to* goods.[18]

The softness of the concept is illustrated by the United Nations' Universal Declaration of Human Rights. Its first articles declare rights of equality before the law, protection from arbitrary arrest, and other classic political and economic freedoms. But later articles go on to declare that "everyone, as a member of society . . . has the right to work, to free choice of employment, . . . to protection against unemployment, . . . to just and favorable remuneration, . . . to rest and leisure . . . and periodic holidays with pay." However worthy these ideals, they remain only proclamations. As with the United States Bill of Rights, manifestoed claims can serve powerfully as ideals and become incorporated into law. The United Nations' Declaration serves as a rhetorical ideal, and some of its provisions have been adopted into countries' legal structures. Much international political attention now is focusing on health rights and the ancillary "basic human needs" rights to healthy birth and infancy, adequate food, and clean drinking water. Still, at bottom, rights-claims have force only as other people accept the correlate obligations they entail (even if only an obligation not to interfere).

Animal rights, rights of other species, and rights of inanimate objects and landscapes are stoutly defended by some.[19] But although I am devoted to protecting wild environments, a great variety of living species, and many landscapes, historic landmarks, and archaeological sites, my devotion does not stem from a belief that these things innately "have rights."

Several leading philosophers continue to promote individual rights as the essence of ethics. At an extreme is Robert Nozick's libertarian view, developed primarily in *Anarchy, State, and Utopia* (1974). He ascribed a great variety of basic property-like rights to individuals, opposed extensive government but defended the need for a minimal state, and suggested procedures for transacting those propertied rights. Nozick viewed rights as historic inherited or earned entitlements that should not be transferred without their owners' consent. He resisted egalitarian prescriptions for redistributing rights and goods, although he did vaguely concede that some compensation might be made to those who are severely disadvantaged. The book strikes me as badly flawed. Nozick did not defend the moral bases of his conception of rights. As for altruism, he would seem only to not forbid it. The simple societies Nozick envisions do not resemble

18. Ronald Dworkin, 1977, is an important essay on legal rights; its sixth printing, 1979, carries a reply to critics. Martin and Nickel, 1980, is a review of standard rights-arguments. Van den Haag, 1983, is a critique of natural-rights theory.

19. Christopher Stone, 1974; Regan and Singer, 1976; Stephen Clark, 1977; Morris and Fox, 1978; Regan, 1982, 1983; and Singer, 1984. 22 *Osgoode Hall Law Journal*, pp. 281–348, 1984, is a symposium on law and ecological ethics.

any postmedieval societies I am aware of. The most I can say of the book is that it has provoked useful debate.[20]

A countertheme through the past several centuries has been the attempt to take account of collective, as opposed to individualist, interests. Philosophers and economists have been striving to formalize their descriptions of the ways people value things, to develop methods for treating values aggregatively (especially those that cannot be exchanged in simple markets), and to develop frameworks that harmonize collective societal with individual interests.

Utility is a notion that helps. By "utility" usually is meant some weighting, or, as I prefer it, transactive valuing, of interests and desires. Jeremy Bentham's utilitarian maxim that society seek "maximum happiness of the greatest number" has spawned offspring more subtle than itself. Scoffers often imply that this rule is all there is to utilitarianism and reject it, and at the same time jettison the whole idea of utility. Most modern decision analyses ask not for such grand summing as called for by Bentham's phrase, but for satisfying people's preferences as expressed in valuing actions. Whereas raw emotive happinesses may be hard to confirm, manifestations of preferences and satisfactions can be solicited and inferred. Selfishness need not be assumed; people's desires may well include a desire that other people be cared for. Nor need utility-oriented cost–benefit analyses, such as those of the banning of lead from gasoline, neglect distributional ethical issues; they can estimate health benefits to children (as reflected, for instance, in medical and educational costs reduced), as well as the increase in fuel costs to automobile operators. Utilitarian decisions tend to seek "rationality," by which is meant prudent, reflective, consistent behavior, ordered proportionately toward specified goals. To the objection that people's relatively uninformed actions aren't reflections of their "true" preferences (what they would prefer in unemotional, rational, fully informed circumstances), analysts respond that they can make the rational adjustments on people's behalf—which of course raises paternalistic issues for those analysts.[21]

The most influential recent attempt to provide an alternative to crude utilitarianism has been John Rawls's *Theory of Justice* (1971).[22] Promoting a Golden Rule, or Kantian, view of justice as "fairness," Rawls attempted to "generalize and carry to a higher level of abstraction the familiar theory of the social contract as found, say, in Locke, Rousseau, and Kant." He would have judgments made from an Original Position, behind a veil of

---

20. Nozick, 1974. Commentary is collected in Paul, 1981.

21. Sen and Williams, 1982, carries articulate arguments for and against various aspects of utilitarianism.

22. Rawls, 1971. For critiques see pp. 150–183 of Ronald Dworkin, 1977; Daniels, 1976; William Galston, 1980; and Sandel, 1982.

ignorance, such that "no one knows his place in society, his class position or social status, nor does any one know his fortune in the distribution of natural assets and abilities, his intelligence, strength, and the like" (p. 12). Justice is, in his account, the first virtue of social institutions, and two principles of justice are fundamental (p. 302):

- Each person is to have an equal right to the most extensive total system of equal basic liberties compatible with a similar system of liberty for all.
- Social and economic inequalities are to be arranged so that they are both to the greatest benefit of the least advantaged, consistent with the just savings principle, and attached to offices and positions open to all under conditions of fair equality of opportunity.

(By "the just savings principle" he means that "each generation must not only preserve the gains of culture and civilization, and maintain intact those just institutions that have been established, but it must also put aside in each period of time a suitable amount of real capital accumulation."[23]) Embracing a form of egalitarian liberalism, Rawls would distribute goods so as to meet everyone's basic life-needs, and would structure distributive actions so as to tend to improve the relative lot of the poorest. As primary goods he lists the basic liberties, freedom of movement and choice of occupation, access to powers and prerogatives of political office, income and wealth, and the social bases of self-respect. He champions constitutional democratic participation as a means for ensuring liberty and justice.

Rawls's book and related publications have been taken very seriously by commentators. Certainly I have found them stimulating. My principal concerns are that Rawls emphasizes justice to the neglect of other moral precepts; that he doesn't cope with the possible conflict between democratic governance (which can favor the better-off) and assurance of greatest benefit to the least advantaged; that he doesn't provide criteria for deciding whether contemplated actions will, in the long run, help the poorest; and that he emphasizes the interests of individuals at the expense of collective societal interests.

It is exceedingly difficult to know how to actually apply this sort of theory in real-world decisions. Several scholars have sought to hybridize Rawlsian with utilitarian or other approaches. Amy Gutmann has proposed a revised Rawlsianism (1980). Professor Rawls continues to refine his theory (1980, 1982).

"If utilitarianism fails to take seriously our distinctness, justice as fairness fails to take seriously our commonality," Michael Sandel has written, in the most perceptive critique I know of on this aspect of Rawls's theory. Sandel urged a constitutive conception, in which citizens think of them-

---

23. Rawls, 1971, p. 285. For commentary on the "just savings principle," see Arrow, 1973.

selves not just as *having* community but as *being* community. "Thus a 'community' cannot always be translated without loss to an 'association,'" he said, "nor an 'attachment' to a 'relationship,' nor 'sharing' to 'reciprocating,' nor 'participation' to 'cooperation,' nor what is 'common' to what is 'collective'" (1982, pp. 150–174).

One other concern. The pursuit of rights and justice, by whatever political philosophy, usually is keyed to "equality." But, as Douglas Rae has correctly warned, "Perhaps the idea against which equality must struggle is equality itself."[24] The profound problem is that there are many kinds of equality. Equal current status (health status) is not the same as equal means or opportunity (access to medical resources), which is not the same as potential for equal attainment (health prospects). As Anatole France's quip that "The law in its majestic equality forbids the rich as well as the poor to sleep under bridges" poignantly made clear, equal treatment may well not provide equal satisfaction. Equal according to needs may not be the same as equal according to wants, and neither is the same as equal according to desserts. Moreover, identical lots are not the same as lots of equal value. Treating individuals (all medical school applicants) equally is not the same as treating blocs of individuals (white male applicants, female Hispanic applicants) equally. Compensating people monetarily for nonidentical treatment is not the same as treating people identically. In an allocation of limited resources, to enhance one person's movement toward some "more equal" status may retard another's. Defining justice as "equal treatment of those equally situated" is different from defining justice as "treatment tending to situate more equivalently in the long run." These distinctions among equalities must be addressed in every specific situation. To call for equality is merely to begin the discussion.

SOME CURRENT ETHICAL CHALLENGES

Briefly now I will survey four current challenges having broad importance to sociotechnical change: questions about the desirability of "naturalness"; about obligations on behalf of future generations; about paternalism; and about protective discrimination.

Is "Natural" Desirable?

To conceive of Nature as a benign order-of-things that somehow "knows best" is to embrace a very elusive abstraction.

Harking back, for reference, to the wild state avails little. Miscarriage is natural. Toothache is natural. Infectious diseases and crop pests are nat-

---

24. Rae, 1981, p. 5; this book instructively discusses some differences among equalities, although not all those mentioned here.

ural. Poison ivy, lice, cockroaches, quicksand, and cholesterol are natural, as are drought, tornadoes, hurricanes, mudslides, lightning, and sleet. Prejudice, selfishness, and jealousy are natural. Surely one of the essential aspects of humanity is *resisting* many natural effects. The increasing realization that natural isn't necessarily desirable is invalidating quite a few traditional ethical touchstones.

No longer can the womb be viewed as a perfect haven for the fetus, for example, or mothers' milk be considered perfect sustenance. Analyses show that a wide variety of potentially toxic heavy-metal ions, pesticides, analgesics, narcotics, natural dietary breakdown products, and metabolic wastes accumulate in the amniotic fluid, making it as much sump as sanctuary. And milk analyses reveal both trace toxic contaminants and lactose and other secretions that can present risks to some nursing infants. A lot depends on the mothers' habits and environment.

Consider pain. Often in the past pain was viewed as natural and, because natural, not to be interfered with. When anesthetics became available for diminishing suffering such as the pain of childbirth, many clergymen and physicians opposed them because they were artificial. Although pain provides useful mental feedback on bodily functions, now in most situations minimization of pain is judged desirable.[25]

As for natural food, what is natural, only foods that grow in the wilderness? In that case, do we abandon agriculture and revert to eating grubs, roots, and berries? Many natural comestibles are potentially toxic (some mushrooms, seafoods carrying toxic microorganisms, peanuts contaminated with aflatoxin), and all are harmful if consumed in excess (fats, starches, constipatives, vitamins, salts). Conversely, most domesticated foods are nourishing when consumed in moderation, and many manmade or processed foods and additives (soyburgers, spoilage retardants, vitamins, reconstituted milk and fruit juices) are beneficial to consumers' health and pocketbooks (Ferrando, 1981). The proper question is not whether foods are natural, but whether they are nourishing.

Humans have intervened in the environment for so long that it is difficult to know what landscapes and ecologies are "natural." Thousands of plants and animals have been moved around the world in the conveyances of man and deliberately or adventitiously "renaturalized"; the chestnut blight, starlings, Japanese beetles, gypsy moths, and countless other alien imports have been pests to the United States, while the eucalyptus trees of California (formerly of Australia), tomatoes (from South America via Europe), and many other imports have of course been beneficial. Of the 1,700 species of plants now considered to be native to Guadeloupe and Martinique, about 1,000 were introduced by man; and many among the 350 of those exotics that have become completely "naturalized" are spe-

---

25. Gibson, 1982, is a good review of pain in cultural context. Also see Pernick, 1985.

cies—coconut, breadfruit, casuarina, mango, banana—that we think of as the very typifying flora of those verdant islands (Fournet, 1977). The heavily grazed hills of Britain and Greece, the cleared American rangelands, the canalled, diked, and drained lowlands of Europe, the rice terraces of the Orient, are unnatural. Game preserves, protected seashores, indeed all farms and managed forests, are thoroughly unnatural (William Thomas, 1956; J. Donald Hughes, 1975; Goudie, 1981). These realizations overwhelm such simplistic prescriptions as the naturalist Aldo Leopold's "land ethic," which argued that "A thing is right when it tends to preserve the integrity, stability, and beauty of the biotic community. It is wrong when it tends otherwise" (1949, p. 224). The tawny hills of Israel, Lebanon, and Jordan, deforested for timber two millenia ago, certainly can be said to have "integrity, stability, and beauty"; but should reforestation there be, on Leopold's principle, refrained from?

One can give many reasons for respecting and preserving wilderness areas, waterways, shores, gardens, and urban parks: for a host of aesthetic and therapeutic pleasures; for quiet communion with the wild and primitive environment from which we organically and historically have derived; for satisfaction of scientific curiosity; and for countless resource and recreation uses. Such delicate and reversible approaches as biological insect control (selectively using ladybird beetles or viruses, for instance, instead of persistent chemicals, to kill agricultural pests) often make good sense. Because we can predict the effects of our interventions only imprecisely, we must respect the intricate "web of nature." But knowledge about wild nature's economies does not provide a sufficient guide for our, concededly, anthropocentric ethics. "While we are *in* nature," René Dubos put it, "we are not quite *of* it" (1973, p. 549).[26]

Health protection regulations sometimes make reference to natural background risk. Automotive ozone emission standards under the Clean Air Act originally were established relative to rural ozone levels (generated by lightning). Ionizing radiation standards are justified in part on grounds that they allow exposure to no more than some specified increment above the radiation received from cosmic rays and geological radioactive decay. This can provide some calibration, but it doesn't help determine how large an artificial increment should be judged tolerable. Background radiation itself induces mutations.

Embodied in many philosophies and policies is the notion of an allotted threescore-years-and-ten natural life-course. The assumption is that everyone should get a fair chance to live to maturity and fulfillment. Anything less is considered premature death, "death before one's time." This ancient deep-rooted ideal is the justification for many protective and com-

---

26. Some authors would disagree: Passmore, 1974, makes a classic plea for "cooperation with nature"; and Rolston, 1982, develops a "redescription of natural value and the valuing process."

pensatory programs. But biomedical life-extension possibilities, and the growing awareness that different genetic constitutions and personal habits make big differences in life expectancies, now are challenging the idea of natural lifespan.

Natural death similarly is problematic. Bioethicist Daniel Callahan has argued that this intuitive concept needs to be redefined, and has proposed the following criteria (1977, p. 32):

> ["Natural death" should be construed as being] the individual event of death at that point in a life span when (a) one's life work has been accomplished; (b) one's moral obligations to those for whom one has had responsibility have been discharged; (c) one's death will not seem to others an offense to sense or sensibility, or tempt others to despair and rage at human existence; and finally (d) one's process of dying is not marked by unbearable and degrading pain.

Some observers have disagreed with Callahan. Dallas High has, for instance, as he also has opposed the 1976 California Natural Death Act, saying that practical disputes over quality of death can better be focused on the right to refuse treatment and other considerations (High, 1978). Invoking naturality conflates *is* (the way people do "naturally" die) with *ought* (the way they therefore ought to be helped to die). Again, it is hard to know how to apply the concept to the high-tech deathbed.

The Behalf of Future Generations

What should be our obligations to our descendants in conserving mineral, land, forest, and other resources? In aiding developing countries? In managing environmental health hazards? In populating the future? In passing on the genes that will become our descendants' genes?

Although these questions may abstractly be of concern to everyone, in technical work they arise concretely, though in broad form. Rarely are the answers either satisfying or specific. We can't know the size or distribution of future populations, what our descendants will need and want, or how technologies will change between now and any future time. Our ancestors of only a few generations ago hardly could have imagined that tuberculosis and smallpox would not be among our principal worries, or that we would care about bowhead whales and California condors and such [then] worthless rocks as pitchblende and bauxite (the ores of uranium and aluminum).

Contractarian theorists extend the notion of justice to cover relations between present and future generations, and rights theorists argue on behalf of rights for the future (MacLean and Brown, 1983; Wenz, 1983). I can't evaluate these arguments here. But I must say that although I agree that we ought to respect future interests, I find it hard to justify this conclusion on mutual-reciprocity contractarian or other strictly reasoned grounds. It seems to me that we should accept our place in the stream of civilization and be guided by an Intertemporal Golden Rule: Act toward future generations as we would have had preceding generations act toward us. A homelier way to put it would be a Campsite Rule: Leave the world no more pressed for resources and no more polluted than we found it, and improved if possible. Thomas Cochran and David Bodde proposed such a criterion for disposal of radioactive wastes: "Nuclear operation of all types (such as mining, milling, fuel processing, decommissioning, and waste isolation or disposal) should be conducted so the overall hazards to future generations are the same as those that would be presented by the original ore bodies utilized in those operations" (1983, p. 117). The distribution of benefits and risks over different social groups is, however, likely to change as time passes, so such a criterion is not easy to apply in practice.

Overall, the problem is that we do, as future generations will, extract virgin natural resources. Conservation, recycling, materials substitution, and development of alternatives will help provide for our successors. We should try to recycle chemical and radioactive materials, minimize the amounts discarded, and dispose of them carefully and reversibly. Some progress is being made in restraining global population growth; indeed some countries have been remarkably successful at controlling fertility. One special concern is that present increases in fecundity, stemming from improved general health and earlier onset of puberty, combined with increased survival of genetically transmitted handicaps because of medical remediation, is increasing the prevalence of those handicaps. Moral discomfort usually has prevented this from being raised as a policy question, but that is likely to change as the problems, and their causes, become more evident.

Paternalism

Science, technology, and medicine long have served as instruments for protecting the health of infants, educating the disadvantaged, and enhancing life for the frail and handicapped. This casts technical experts, as the agents wielding those instruments, into a variety of paternalistic roles. Increasingly, paternalistic policies of government and industry, and paternalistic actions on the part of professional experts, are being challenged.

A sense in which paternalism is viewed as bad is embodied in such definitions as Gerald Dworkin's: "By paternalism I shall understand roughly the interference with a person's liberty of action justified by reasons referring exclusively to the welfare, good, happiness, needs, interests or values of the person being coerced." As examples Dworkin listed laws that require motorcyclists to wear safety helmets, laws forbidding women and children to work at certain types of jobs, civil commitment procedures justified on grounds that they prevent the person being committed from harming himself, and fluoridation of public water supplies.[27]

Dworkin's definition strikes me as being much too narrow, stressing the coercive aspect of paternalism more than does common usage, which implies, more broadly, "dealing with individuals in the manner of a parent." John Rawls gave paternalism more latitude than Dworkin did (1971, p. 249):

> Others are authorized and sometimes required to act on our behalf and to do what we would do for ourselves if we were rational, this authorization coming into effect only when we cannot look after our own good. . . . We must be able to argue that with the development or the recovery of his rational powers the individual in question will accept our decisions on his behalf and agree with us that we did the best thing for him.

In a medical context, many would find Allen Buchanan's definition useful: "Paternalism is the interference with a person's freedom of action or freedom of information, or the dissemination of misinformation, or the overriding of a person's decision not to be given information, when this is allegedly done for the good of that person."[28]

The problem is that, like parental acts, most paternal acts are driven by both altruism and self-interest, and they may be intended as much for the collective good of society as for the good of the affected individuals.

Rarely do people disapprove of actions to protect fetuses, infants, children, the mentally unwitting, and the senile, against bodily or emotional harm. Americans expect technical experts and the government to protect them against hazards, even self-inflicted ones, about which they may not be fully aware, such as harmful foods, cosmetics, pharmaceuticals, toys, and complex technological hazards. But on matters about which people feel competent to make their own decisions, such as seat belts, highway speed, saccharin, consumption of antibiotics, and cigarette smoking, they often challenge authorities' paternalism. As has been true ever since Tocqueville observed it, "Our contemporaries are ever prey to two conflicting passions: they feel the need of guidance, and they long to stay free."[29]

The laetrile controversy is a typical issue. Under the federal law that requires the U.S. Food and Drug Administration to forbid pharmaceuticals to be sold in interstate commerce unless they have been proved both safe and efficacious, the agency banned laetrile as being of doubtful effi-

27. Gerald Dworkin, 1971. Dworkin, 1983, presents slightly tempered second thoughts.

28. Buchanan, 1983, p. 62. Regarding the delegation of authority over one's irrationalities, see Elster, 1979.

29. Tocqueville, [1835] 1966, p. 667. On the philosophic grounds of paternalism, see Lively, 1983; Sartorius, 1983; and Kleinig, 1983.

cacy against cancer. But some cancer patients convinced otherwise, desperate for a cure, and resentful of this unwanted paternalism, have successfully lobbied quite a few state governments to allow preparation and sale of laetrile within those states (Markle and Peterson, 1980).

It is the essence of medical practice that doctors paternally "do what is best for the patient." In routine care this may be straightforward—although even then it may involve such tactics as deception (as when, to take a simple case, a placebo is administered, deceptively by definition, to enhance a patient's expectation of cure and thereby enhance cure). The assumption is that on becoming "a patient" the client surrenders some autonomy and grants some license to the physician. Medicine enacts stronger paternalism when it intervenes on behalf of the welfare of persons judged not fully competent to make their own choices or care for themselves, or who are at risk of harming themselves or others.[30]

For psychiatrists and psychological counselors the problem is to decide, in each client's case, how paternal to be and, indeed, what paternal might mean: whether to urge correction toward "more normal" behavior, or to help the client attain adjustment and satisfaction in his abnormality. There is marked disagreement within the profession. The issue was raised clearly by one interview study of psychotherapists (Buckley et al., 1979, p. 218):

> Almost all of the 81 respondents agreed that the therapist should not impose his value system on the patient, yet half of the group viewed the therapist's encouragement of the enrichment of the patient's social life and encouragement of educational and vocational pursuits as an important aspect of the therapy. A significant number agreed that giving suggestions and advice to the patient may be harmful, but an equal number felt that the therapist should encourage more adaptive modes of behavior, including sexual intimacy in the patient's outside life.

In years coming, scientific and medical paternalism will be contended as health experts emphasize such "clean living" health factors as diet, exercise, and moderation in drug consumption (just as their expert predecessors, controversially, campaigned against vitamin and mineral deficiencies). We should expect major arguments over protection of human sperm, ova, and developing fetuses. Already there has been a case of a physician's taking a pregnant woman to court to try to force her to stop taking Quaalude, Valium, cocaine, and morphine, in the interest of the fetus.[31]

---

30. On medical paternalism, see Veatch, 1981, pp. 195–213; and Beauchamp and Childress, 1983, pp. 168–182.

31. "Pregnant Drug User Taken to Court," *New York Times*, April 27, 1983; the lawsuit was dropped, so legal dimensions remain unclear.

Protective Discrimination

Related to paternalism is the issue of protective discrimination. I would argue that "discrimination" need not be viewed as pejorative. We willingly select people differentially for jobs, military tasks, athletics, leadership, and other functions. Bodily constitutions differ widely. Allergies are common and staggeringly diverse. Genetic variance among individuals is large. As we learn more about immunological, genetic, and other basic characteristics of people, fundamental differences will become ever more evident.

In the workplace, almost certainly we will want to discriminate between older and younger, the strong-backed and the weak-framed, potential childbearers and those past childbearing, potential fathers and those past fathering, the overweight stroke-prone and the cardiac robust, smokers and nonsmokers. This will conflict directly with equal-job-opportunity and other values.

Questions now are being raised over whether genetic screening should be a condition for chemical industry employment. Hypersusceptibility to intoxication or dermatitis can be genetic (for example, glucose-6-phosphate dehydrogenase deficiency, which is genetically intrinsic in quite a few people, increases the risk of acute hemolysis from some occupational chemicals).[32] Health screening of any kind raises mixed prospects. On the present issue Thomas Murray, while not clearly disapproving of genetic testing, has argued:[33]

> Our genetic makeup is fixed, unchangeable, and unchosen. It is also inherited along racial and ethnic lines. It is out of our control, and all the more frightening because of that. It raises at times the specter of racial discrimination. Put another way, genetic tests are nakedly antiegalitarian.

But the counterargument is, I believe, at least as persuasive: that failure to take into account differential susceptibilities and provide equivalent, though possibly not identical, protection for all workers—and their incipient offspring—would be even more antiegalitarian than screening is. Until that impossibly far-off day when perfect occupational protection is available for all, equal protection almost certainly will require assigning workers discriminately. A recent study by the congressional Office of Technology Assessment framed the ethical questions:[34]

32. Omenn, 1982, is a medical descriptive review. Lappé, 1983, essays on the ethics.

33. U.S. House of Representatives, Subcommittee on Investigations and Oversight, of the Committee on Science and Technology, October 1982, p. 76.

34. U.S. Congress, Office of Technology Assessment, 1983, p. 141.

- What moral rights and duties exist between the worker and company medical personnel?
- Must participation in genetic testing programs be voluntary, and if so, how is that to be guaranteed?
- What rights and obligations exist regarding the use of medical information?
- What ethically permissible actions may be taken on the basis of information gained through genetic testing programs?
- Do the answers to these questions depend on whether the testing is being done for research purposes or as part of a medical program?

Analogous tribulations are arising over control of worker exposure to reproductive hazards: whether to treat women of childbearing age differently from men; whether to forbid men to work at possibly spermatotoxic jobs; how to treat pregnant women [and their babies in utero]. And now that environmental protection has been established for average citizens, we must move toward the uncertain margins and confront the troubling questions of how to protect those who are unusually vulnerable.

BRIDGING FROM IDEALS TO PRAGMATICS

Formal principles and goals can help relate practice to ideals.

Bioethical Principles

In medicine, ethical principles have been promulgated for several millenia. But now that so many factors that affect health are seen as lying outside of traditional medical boundaries, and now that the context of health has so clearly become economic and political, those guides are having to be reformulated. Because of their experiential immediacy and poignancy, biomedical issues tend to serve as forerunners for other technical issues.

Bioethical principles proposed usually include *respect for "the person" and personal autonomy*, *nonmaleficence* (the *primum non nocere*, "above all, do no harm," precept of classical medicine), *beneficence* (helping, caring, offering compassion and charity), and *pursuit of justice*.[35] Principles of this kind now are being adopted in technical arenas outside of standard medicine, such as in guidelines for genetic counseling and genetic alteration, in regulations to protect subjects of social-science research, and in environmental health standards.

How hard it is to relate such general precepts to complex real choices, even in essaying one step toward specificity, is evident in such attempts as

---

35. Beauchamp and Childress, 1983, is a text carefully developed around these principles. Shelp, 1982, is a symposium on beneficence. General sources include Ramsey, 1978; Veatch, 1981; and Pellegrino and Thomasma, 1981.

the following, by Robert Veatch. "It is a mistake to base a public policy on any estimate of the value of life years in the aggregate or abstract," Veatch argued. Rather, policies "ought to focus on ethical claims that cannot be captured in such estimates: on the fairness of the distribution of life years, on the quality of the life years with special attention to the relief of suffering, and on whether disease and death that shorten lives are seen as in some sense the responsibility of the individual or whether disease and death are seen as beyond individual control." On these precepts Veatch proposed guidelines for choosing among life-extending technologies:[36]

- The relative social usefulness of individuals or members of a class, their willingness to pay for life-extending technologies, and the dollar value of their future production ought to be irrelevant to policy choices regarding alternative life-extending technologies.
- Any other aggregating methods of determining which life-extending technologies deserve priority should be used cautiously, if at all.
- The younger the individual, the greater should be the priority of life-extending technologies.
- Medical conditions that produce the greatest suffering should be given priority.
- The needs of the generally least well off at times may have to be placed ahead of those who may die relatively early because of a medical problem.
- A medical condition that is seen as involuntary, which originated through causes outside the control of the individual, should get priority over conditions which have resulted from voluntary choice of health-risky behavior and life-style.

Although each of these considerations is endorsed in some contexts, I doubt that any of them would find consensus approval. And certainly it is asking a lot to propose the cluster as universal criteria.

Nevertheless, criteria of this general kind are exactly what we must formulate. We can't do without them. My own view is that we need to integrate considerations of degree of suffering, voluntariness of riskbearing, helplessness, and so on, with the pragmatic considerations of premature-death analyses, health-status-preference indices, and quality-of-life-adjusted risk–cost–benefit appraisals.

Goal-Setting

Operational goals, such as injury-reduction goals in a steel mill, are a way of yoking practice to intentions, as are such national objectives as

36. Veatch, 1980, p. 158.

"achievement of fishable and swimmable rivers." But the more general the intentions, and the more complex the operations, the more important but at the same time the more difficult it is to establish goals that provide practical guidance.

Much pragmatically influential goal shaping occurs in national legislation and jurisprudence. Laws address such goals as the development of international communications satellite systems, the fostering of urban renewal, and radiation protection. Over the years these goals get refined, as parliamentary or legislative intentions are reviewed, as the public reacts, as conflicts among goals get sorted out, as budgets are appropriated. Always the problem is to devise specified, applicable goals.

In 1979 the U.S. Surgeon General issued a report on health promotion and disease prevention, *Healthy People*, which reviewed the remarkable public health gains of this century, described the principal current health threats, and urged a recasting of health strategy toward prevention. The report proposed a set of ambitious general goals for 1990, such as "to reduce infant mortality by at least 35 percent, to fewer than nine deaths per 1,000 live births." A sequel, *Promoting Health/Preventing Disease: Objectives for the Nation*, "the result of a year long effort involving more than 500 individuals and organizations from both the private and governmental sectors," detailed many quantified objectives in pursuit of those goals, such as "By 1990, the proportion of women in any county or racial or ethnic group . . . who obtain no prenatal care during the first trimester of pregnancy should not exceed 10 percent. (In 1978, 40 percent of black mothers and 45 percent of American Indian mothers received no prenatal care during the first trimester. . . .)" As of early 1984 these objectives were said to be the Public Health Service's "highest priority in prevention," and federal agencies were implementing plans for attaining those objectives. These goals and plans are to be reviewed periodically.[37] Is this process really working?

Concerned about the long-range energy future, Amory Lovins has urged that we follow "soft" energy paths. Characteristics that make paths soft—"not vague, mushy, speculative, or ephemeral, but rather flexible, resilient, sustainable, and benign"—are (1977, p. 38):

- They rely on renewable energy flows that are always there whether we use them or not, such as sun and wind and vegetation: on energy income, not on depletable energy capital.
- They are diverse, so that as a national treasury runs on many small tax contributions, so national energy supply is an aggregate of very many individually modest contributions, each designed for maximum effectiveness in particular circumstances.

37. U.S. Office of the Assistant Secretary for Health and Surgeon General, 1979, 1980, 1983.

- They are flexible and relatively low technology—which does not mean unsophisticated, but rather, easy to understand and use without esoteric skills, accessible rather than arcane.
- They are matched in *scale* and in geographic distribution to end-use needs, taking advantage of the free distribution of most natural energy flows.
- They are matched in *energy quality* to end-use needs.

Debating and adopting criteria of this form (of whatever content) can help conform technical programs to values.

A 1980 Rand Corporation study, *Issues and Problems in Inferring a Level of Acceptable Risk*, described how energy programs can be related to various risk-reduction goals or criteria for choice, such as minimization of maximum accident consequences, minimization of probability of most probable accident, and so on. After describing how goal choices can make a difference to programs, the report urged that energy decisionmakers "be *self-aware* in specifying risk-reduction goals, as well as in relating them to goals of other agencies and interested parties, and understanding their implications for the choice of energy alternatives" (Salem, Solomon, and Yesley, 1980).

But nuclear power regulation illustrates the difficulty in doing this. The U.S. Nuclear Regulatory Commission, seeking to develop guidance on nuclear power plant siting, design, operation, maintenance, and inspection, long has struggled with whether it should translate the legislated mandate of "adequate protection of public safety and health" into specific objectives. In 1983 the Commission adopted a "Policy Statement on Safety Goals for the Operation of Nuclear Power Plants," a controversial proposal that will be subjected to formal review at least through 1985. The policy began by stating two overall goals:

- Individual members of the public should be provided a level of protection from the consequences of nuclear power plant operation such that individuals bear no significant additional risk to life and health.
- Societal [summed] risks to life and health from nuclear power plant operation should be comparable to or less than the risks of generating electricity by viable competing technologies [coal especially] and should not be a significant addition to other societal risks.

Then it went on to specify accident prompt-fatality limits (0.1% of the national fatal-accident burden) and cancer-fatality limits (0.1% of the national cancer risk); to propose a benefit–cost guide for safety improvements (suggested limit: adopt any improvement that costs up to $1,000 per marginal person-rem of radiation exposure averted); and to set a limit on large-scale core-melt accidents (no more than one such accident per

10,000 reactor-years of operation).[38] The Commission proposed these goals only after considering and rejecting quite a few other formulations and limits. But the goals remain controversial. Some critics are contemptuous of the entire approach, believing that the totality of engineering judgments embodied in standards and practice are better guides than any grand goals are. Others quarrel with the particular limits proposed. Others find the goals of little help in making real "nitty-gritty" hardware decisions. Still others argue against probabilistic analysis, and thus against goals based on it, altogether. I strongly recommend review of the exercise to anyone, in any area, trying to couple technical action with value-explicit goals.

38. U.S. Nuclear Regulatory Commission, 1983, sets forth the proposed goals and prints many reviewers' commentaries.

Nine

Science and Technology in the *Polis*

In his 1882 play *An Enemy of the People* Henrik Ibsen presented with dramatic compression the issues we must address now.[1]

MORTEN KIIL [a neighbor]: Let me see, what is the story? Some kind of beast has gotten into the water pipes, is it?

DR. STOCKMANN [physician]: Bacteria, yes.

MORTEN KIIL: And a lot of these beasts have gotten in, according to Petra; a tremendous number.

DR. STOCKMANN: Certainly; hundreds of thousands of them, probably.

MORTEN KIIL: But no one can see them—is that right?

DR. STOCKMANN: Yes; you can't see them.

MORTEN KIIL (with a quiet chuckle): Damn—it's the finest story I have ever heard!

DR. STOCKMANN: What do you mean?

MORTEN KIIL: But you will never get the mayor to believe a thing like that. . . .

PETER STOCKMANN [mayor, and the doctor's brother]: Have you taken the trouble to consider what your proposed alterations would cost? According to the information I obtained, the expenses would mount up to fifteen or twenty thousand pounds. . . . I have not, as I remarked before, been able to convince myself that there is actually any imminent danger. . . . Nothing of this unfortunate affair—not a single word of it—must come to the ears of the public. . . .

---

1. My translation.

DR. STOCKMANN: Well, but is it not the duty of a citizen to let the public share in any new ideas he may have?

PETER STOCKMANN: Oh, the public doesn't require any new ideas. The public is best served by the good-old-established ideas it already has. . . . What we shall expect is that, after making further investigations, you will come to the conclusion that the matter is not by any means as dangerous or as critical as you imagined in the first instance. . . . As an officer under the [Public Baths] Committee, you have no right to any individual opinion.

DR. STOCKMANN: No right?

PETER STOCKMANN: In your official capacity, no. As a private person, it is quite another matter. But as a subordinate member of the staff of the Baths, you have no right to express any opinion which runs contrary to that of your superiors.

Details comprehensible only to experts; . . . but consider the cost; . . . the danger may not really be imminent; . . . the public needn't know; . . . no right to challenge the wisdom of superiors, anyway. . . . In the hundred years since *Enemy* only moderate progress has been made on the issues Ibsen raised, and many new subplots of expertise and role conflict have been revised into the drama.

A present-day Stockmann still would have to contend with fact/value issues, perform multiple roles, and agonize over how vigorously to press his advocacy. But in a way that his centennial predecessor never could have, now he would be able to tap a large legacy of professional judgment, such as that woven into National Research Council Safe Drinking Water Committee studies and Environmental Protection Agency regulations. And in repressive circumstances, if he were aware of the recourses, he could call upon guidance and support from microbiology, public health, and sanitary engineering colleagues and professional organizations.

EXPERT ADVISING, ADVOCATING, AND WHISTLEBLOWING

Three generic questions surround the activities of advising, advocating, and whistleblowing. First, what mode of knowing and judging is involved? Second, in their rendering of opinion in the case at hand, what social roles are the various experts choosing to play, and what roles are they playing, de facto? And third, what assumptions, premises, biases, and intentions underlie the experts' views—in current vernacular, "Where are they coming from?"

A prototype of the first issue, that of mode of knowing and judging, was the question Prime Minister Churchill asked Lord Cherwell in 1944 after Niels Bohr had briefed them on the prospects for international control of atomic energy: "What is he really talking about? Politics or physics?" Physics *in* politics, no doubt, or politics surrounding physics (Jungk, 1958, p. 174). Technical-fact/social-value distinctions are typical issues in such queries. So are the grounds of technicolegal certifications, such as a doctor's attestation that a person has competently consented to being experimented on (a "medical" judgment?).

The second issue, that of roles, was raised, in almost caricature form, in the 1971 Senate hearings on supersonic aircraft development. Senator Gordon Allott harassed professor George Rathjens of the Massachusetts Institute of Technology:[2]

> SENATOR ALLOTT: I would like to ask you a couple of questions, Doctor. As I understand it, you were trained as a chemist. Is that correct?
>
> DR. RATHJENS: That is correct.
>
> SENATOR ALLOTT: And you are now a professor in political science.
>
> DR. RATHJENS: Yes, I am.
>
> SENATOR ALLOTT: Are you testifying today as a political scientist or as a chemist?
>
> DR. RATHJENS: I would not say, Senator, that I am testifying as either. I am testifying as a concerned citizen. I have looked at the problem as best I could, taking into account the combination of largely technical considerations and economic considerations.
>
> SENATOR ALLOTT: Well, you just stated that you were not an authority on economics. You state now that you are not testifying either as a chemist or a political scientist, and I take it you are just testifying, then, on a broad, general feeling as to how you feel about this thing.

Rathjens went on to serve ably in the hybrid role of citizen and technology policy expert.

Speaking as a technical specialist; speaking as a policy expert; speaking as an Aesculapian Notary; speaking as a citizen. Although these kinds of roles can be distinguished to an extent, often they overlap so much that it is artificial to try to split them. Besides, the experts themselves may insist that they are speaking in several roles. The American Psychiatric Association's statement, published after the courts found John Hinckley "not guilty by reason of insanity" of trying to assassinate President Reagan, began: "The American Psychiatric Association, speaking as citizens as well as psychiatrists, believes that the insanity defense should be retained in some form" (American Psychiatric Association, 1982). (Actually, since one assumes that experts always speak "as citizens," the question to the psychiatrists is in what sense they are speaking as psychiatrists rather than as legal philosophers.)

Since everyone plays many roles, even in public forums, there is much room for ambiguity and confusion. Any expert serving on a government

2. U.S. Senate, Committee on Appropriations, 1971, p. 472.

advisory committee or testifying in a public inquiry typically will, among many other things, be a "member of a profession" in some sense (which, as we have seen, takes different forms in different professions), be an employee of some institution, and have some political identity. So then, if a Republican licensed as a chemical engineer, working as an Exxon environmental manager, and serving on committees of the American Petroleum Institute and the Chemical Manufacturers Association, testifies before a congressional hearing on chemical waste, in which roles is she speaking? If a Ph.D. geneticist on the staff of the Natural Resources Defense Council serves on an Environmental Protection Agency advisory committee, can he escape being viewed as speaking on behalf of his employer? And what intellectual fidelity does he owe the (rather vague) professional standards of that amorphous and unlicensed body, geneticists? We must move beyond condemning experts as being compromised-by-affiliation. Relevant experts must be tapped wherever they reside, and their scientific statements held to rigorous scientific criteria. Better to counter any analytic or prescriptive "cheating" with scientific counterarguments, rather than with accusations of association.

When experts extend the reach of their wisdom far beyond usual boundaries, questions arise that resemble those about baseball players advertising shaving cream. Internal logical merits of their arguments apart, should doctors calling themselves Physicians for Social Responsibility be accorded any deference when they claim they speak *as physicians* in taking stands on nuclear weapons? Since economic policy also affects people's health, why shouldn't these M.D.'s pronounce on income taxes, or steel import quotas? Remembering an example from the outset of this book, how does the nuclear-policy legitimacy of the doctors' group differ from the foreign-policy legitimacy of Artists Call Against U.S. Intervention in Central America? In the sourcebook for health professionals, *Nuclear Weapons and Nuclear War*, Christine Cassel and Andrew Jameton justified involvement (1984, p. 9):

> We believe that there is a forceful argument for medical responsibility with regard to preventing nuclear war, and that this argument withstands common objections. The argument can be outlined as follows:
>
> 1. Physicians have a special and central professional responsibility to treat disease and to reduce mortality.
> 2. A large-scale nuclear war would cause death and illness on a massive and unprecedented scale.
> 3. Physicians would be unable to intervene effectively in the human injury and death expected in a large-scale nuclear war.
> 4. Prevention is the only way to reduce mortality where treatment is ineffective.

5. A large-scale nuclear war is possible, or even probable, in the decades ahead.
6. Efforts by physicians could help prevent nuclear war.
7. Therefore, physicians have a central and urgent professional responsibility to help prevent nuclear war.

Almost needless to say, not everyone, and certainly not every physician, is swayed by this reasoning.

In any advisory or advocacy situation, experts should declare what roles they are playing. And when they speak on larger social issues—which from time to time they should be encouraged to do—they should be candid about the degree of stretch in their reach.

The third set of generic issues, that of experts' underlying assumptions, premises, biases, and intentions, has been discussed in every chapter of this book and need not be critiqued here. Many can be inquired into in commonsense fashion, and some can be explored more formally. The most important thing is to maintain sensitive awareness of their influence.

Advising

Uncountable thousands of scientific experts serve as advisors to government and the military, to industry, labor, and other special-interest organizations, and to foundations, hospitals, and museums. Advising conveys knowledge from those who generate and nurture knowledge to those who make decisions and exert social power. That conveyance needs to be grounded in a technical community matrix and certified by critical peer review.

How is expertise recognized? Recalling Chapters Three, Four, and Five: by the technical peerage's rating of a person's substantive accomplishments, including earned proxies of accomplishment such as grants, awards, and appointments, and by its evaluation of the person's record in advising, testifying, and leading sociotechnical endeavors. Recognition of expertise remains social. Failure to appreciate this leaves one holding the absurd view that all advisors are equivalent, that reliable knowledge can't be expected, and that uncertain science can only be dealt with by counting votes from everyone who expresses an opinion.

Of course, being certified a technical expert no more guarantees correctness than being acknowledged an expert cook guarantees perfect meals. But as Bertrand Russell said in his *Sceptical Essays* (1928, p. 13):

> The opinion of experts when it is unanimous must be accepted by non-experts as more likely to be right than the opposite opinion. The scepticism that I advocate amounts only to this: (1) that when the experts are agreed, the opposite opinion cannot be held to be certain; (2) that when they are not agreed, no opinion can be regarded as cer-

tain by a non-expert; and (3) that when they all hold that no sufficient grounds for a positive opinion exist, the ordinary man would do well to suspend his judgment.

Any use of advisors involves a cluster of perennial concerns. In the appointment of advisors, should experts' partisan political affiliations be a consideration? How can conflicts of personal interest be avoided, or at least tempered? Should advisors be made privy to confidential information but sworn to loyalty and confidentiality, thereby ensuring best-informed advice and close relations with those they advise, or should they be kept "outside" but left free to criticize? Should advisors be chosen who are deep but narrow specialists, or should generalists be preferred, for their breadth and perspective, even though they may not be masters of the intricacies? Usually a diversity of advisors and advisory channels is tapped.[3]

Through the course of this book I have made many suggestions on advising: that advisory charges be clearly understood between advisees and their advisors; that potential for financial, political, or personal conflicts of interest be placed on open record; that value-laden judgments be addressed explicitly and candidly; and that meta-analytic issues, such as those having to do with uncertainty and indeterminacy, be dealt with carefully.

An indispensable species is the expert generalist, the person having a reputation for technical soundness, skill in shaping inquiries, the savvy to integrate analysis with prediction and advice, the leadership needed to induce committee and staff members to serve effectively, and the stamina required to press on through controversies. Like the mathematical masters I lauded earlier, this species is rare.

Some specialists grow into generalists informally. One problem this raises, though put a little too self-effacingly, was expressed by Jerome Wiesner: "I've been billed as an expert on arms control, and I think I'm an example of what's wrong with the American posture in this field. I've become an expert in arms control by spending a small fraction of my time since 1958 working on the problems of arms control. . . . My background is primarily in the field of military technology."[4]

Filtering, shaping, and interpreting advice for decisionmakers is a high administrative craft. Richard Neustadt and Harvey Fineberg have reviewed the Ford administration's ill-fated 1976 decision to try to innoculate Amer-

3. Nichols, 1972, and Mazur, 1973 provide general commentary on technical expertise. Beckler, 1974, reflects on twenty years of White House science advisory staff service. Golden, 1980, collects commentaries on science advice to the President. U.S. Congress, Congressional Research Service, 1979, compiles thorough case studies of how technical information has been provided to the Congress. Aaron, 1978, discusses social-science and policymaking expertise in the 1964–1968 Great Society era.

4. Wiesner, 1961, p. 173. Gilpin and Wright, 1964, discusses the problem of scientist-as-strategist.

icans against the epidemic of swine flu that, as it happened, did not occur.[5]
In conclusions that could be drawn from many other such episodes, the
authors found the swine flu affair to have been marked by "overconfidence
by specialists in theories spun from meagre evidence; conviction fueled by
a conjunction of some preexisting personal agendas; zeal by health professionals to make their lay superiors do right; premature commitment to
deciding more than had to be decided; failure to address uncertainties in
such a way as to prepare for reconsideration; insufficient questioning of
scientific logic and of implementation prospects; and insensitivity to media
relations and the long-term credibility of institutions."

> Health was an absolute value. This is as simplistic as [the] next step,
> bundling up all pieces of the issue into one decision with one deadline,
> and pressing it on [the Secretary of Health, Education, and Welfare
> (HEW)]. "Strong possibility, with probability unknown," once down
> on paper, leakable at will, is at an opposite extreme from the detailed
> definition of relevant assumptions we suggest that a decisionmaker
> seek. . . .
>
> Among the things Ford was *not* warned about were six: trouble with
> serious side effects, with children's dosages, with liability insurance,
> with expert opinion, with Public Health Service public relations, and
> with his own credibility. On the contrary, the vaccine was presumed
> both safe and efficacious; insurance was not known to be a problem;
> experts were pronounced on board; [HEW officials] could cope with
> the press.

It is difficult to know how to affix culpable responsibility on advisors. An
advisee can, of course, judge whether his advisors have filled their basic
duties. But many advisors, such as those serving on the Science Advisory
Board to the Environmental Protection Agency, advise not only the agency
administrator but, in an important sense, the American public. Ethical criteria for holding advisors responsible, Dennis Thompson has suggested,
include: whether the advising relates causally to consequences; whether
the advisors can realistically be expected to foresee consequences that flow
from their advice, and whether they intend those results; and whether any
harm that occurs stems from failure of the advisors to challenge role-
bound constraints on their advising (1983).

There is one ironic need: protection against the disincentives that arise
as scientists improve their predictive abilities. As seismologists get better
at foreseeing the location, time, and force of earthquakes, they are urged
to warn the public of disaster. Some are employed or retained as consultants by the U.S. Geological Survey and other authorities to stand watch.
Yet the forecasting craft is still very imprecise and will always be probabi-

5. Neustadt and Fineberg, 1983, pp. 88 and 24. Also see Silverstein, 1981.

listic. Geologists would like to prepare to issue warnings. But they fear that if they call a major alarm and cause citizens to close businesses, cut off gas and electricity, and evacuate, but the earthquake then fails to occur, the predictors will be severely penalized. They will equally be at jeopardy if they fail to warn and the earthquake does occur. "Unless some legal protection is forthcoming," one geologist has warned, "the matter will most likely be resolved by the scientists retreating to an uninformative conservatism when asked to interpret their results in the public forum."[6] Toxicologists, meteorologists, and other predictors face the same predicament.

Advocating

Beyond advising quietly, technical leaders often play advocacy roles, pressing their convictions on society. It is quite a legacy (Chapter Two mentioned other examples). Jenner and Pasteur vigorously pursued vaccination, and Lister antisepsis. From the 1850s on, sanitary engineers campaigned for municipal waterworks, sewers, and garbage collection (Schultz and McShane, 1978; Melosi, 1980). Soil conservationists sermonized against land despoilation (Batie, 1983). In the nutrition revolution of the 1930s, biochemists sought vitamin fortification of milk and bread (National Research Council, 1944; Wilder, 1956). At the close of World War II many scientists crusaded for international control of the atom (Alice Kimball Smith, 1965). Geneticists and others opposed atmospheric testing of nuclear weapons (Kopp, 1979). During the 1950s and 1960s Jonas Salk, Albert Sabin, and their colleagues strongly advocated polio vaccination (John Paul, 1971). With *Silent Spring* Rachel Carson spurred reexamination of DDT and other chemical pesticides.[7] In the late 1960s groups such as the American Association for the Advancement of Science pursued concern over health and environmental effects of Agent Orange in Vietnam.[8] Paul Ehrlich raised public consciousness with his *Population Bomb* (1968). Surgeon generals and their advisors have promoted drinking water fluoridation and have campaigned against smoking. With support from the Rockefeller Foundation, Norman Borlaug led the so-called Green Revolution. Irving Selikoff has fought asbestos. Linus Pauling has pushed vitamin C. A group called Concerned Coastal Geologists has been arguing for radical change in shoreline protection policy ("Along U.S. Coasts," 1981; Kerr, 1982).

In advocacy, experts convey their opinions with force of commitment and crusading. The vigor of that crusading is what distinguishes advocacy from advice.

6. Raleigh, 1981. Also see NAS/NRC, Panel on the Public Policy Implications of Earthquake Prediction, 1975.

7. Carson, 1962. For commentary, see Lowrance, 1976, pp. 155–173.

8. U.S. Congress, Congressional Research Service, 1979, pp. 591–652.

Obviously advocacy in itself is neither good nor bad. Although some is mere grandstanding or establishment-baiting, our lives are enriched by much of it. When advocacy is driven by paternalistic motives, as with anti-smoking campaigns, it may be resisted by publics who resent being paternalized. When it concerns such prosaic commonweal goods as soil, or mangrove swamps, or helium reserves, it may find passionate public interest hard to generate. Technical institutional stewardship, to which I will come shortly, holds a key to its quality and legitimacy.

Some commentators believe that since perfectly dispassionate advising is impossible, on policy-related problems it is better for experts openly to take an advocacy approach. Alice Rivlin, for instance, has encouraged the development of "forensic social science, . . . in which scholars or teams of scholars take on the task of writing briefs for or against particular policy positions." Reviewing Christopher Jencks's controversial book *Inequality*, she explained (1975, p. 62):

> In judging the Jencks book as an example of forensic social science, one should not ask: Is it balanced and objective? Rather, one should ask: Do Jencks and his associates make clear the position they are trying to prove or disprove? Is their evidence convincing? Do they present the evidence fully and fairly? Have they recognized and dealt with the major findings by other scholars which appear to run counter to their position? In other words, do they abide by the high standards we would all like to see set by social scientific advocates in the public policy court?

This approach deserves exploration not only in social-scientific but in all technical aspects of policymaking.

More and more, technical people are being called on to render expert legal testimony. Anthropologists are testifying on such questions as American Indians' cultural interpretation of hunting, fishing, and whaling, as has been raised, for example, in international whaling negotiations, and the religious significance of traditional burial grounds, as brought up in several nuclear powerplant and liquefied natural gas facility siting debates.[9] Radiologists, pathologists, psychiatrists, and other physicians testify in medical cases. Toxicologists and epidemiologists serve as witnesses in court and regulatory proceedings. The troublesome questions are how courts and regulatory agencies should distinguish highly qualified technical witnesses from those less qualified, recognize the difference between analysis and advocacy, and assess the factual and valuative correctness of advocacy testimony.[10]

---

9. Rosen, 1977, and Delgado and Millen, 1982, are reviews.

10. Brent, 1982, criticizes unqualified and irresponsible medical testifying. Cohen, Ronen, and Stepan, 1980, is a bibliography of science–law issues in general.

Occasionally advocates become "true believers" in a cause, get passionately involved in their fight, and exceed the bounds of science or civility. Anyone with minimal expert standing who wishes to create havoc can do so, and may gain amplification in the newsmedia. The public becomes confused in trying to choose among the various self-pronounced and established expert opinions. In Western countries, technical authorities are not set up in a firm hierarchy (which, in many ways, is a strength). Even such eminent institutions as the National Academy of Sciences are not held to be impeccable. And sometimes, of course, mavericks turn out to be right. They may even be correct in the broad, as Rachel Carson was, despite being incorrect in some particulars. The history of science is, after all, a history of heterodox opinions becoming orthodoxy. What we need are ways of hearing advocates out, but at the same time keeping the quality of discourse high and the tone temperate.

Whistleblowing

An especially intense mode of advocacy is whistleblowing. In the last decade this form of principled dissent, which by definition occurs within institutional contexts, has become commonplace. Although few whistles actually are blown, the notion of whistleblowing has become familiar, and many institutions have developed routines for accommodating constructive dissent.[11] In technical arenas, such dissent may concern carelessness, incompetence, or fraud in research or service, improprieties in rendering broad sociotechnical judgment, or outright misfeasance.

Typical of the codified bases for whistleblowing is the Code of Ethics of the National Society of Professional Engineers, which stipulates:[12]

> Engineers shall at all times recognize that their primary obligation is to protect the safety, health, property, and welfare of the public. If their professional judgment is overruled under circumstances where the safety, health, property, or welfare of the public are endangered, they shall notify their employer or client and other such authority as may be appropriate.

Engineers have blown the whistle on substandard nuclear reactor pipe welding, on faulty DC-10 door and hydraulic control design, and on managerial lapses in the development of the Automatic Train Control system for the San Francisco Bay Area Rapid Transit trains.[13] Biomedical

11. U.S. Senate, Committee on Governmental Affairs, 1978, reports some federal government cases. Westin, 1981, provides an overview. Glazer, 1983, describes some whistleblowers' experiences.

12. National Society for Professional Engineers, 2029 K Street NW, Washington, DC 20006.

13. Unger, 1982, analyzes cases. Anderson et al., 1980, chronicles the BART episode and shows it to be less simple a case than is sometimes portrayed. Weil, 1983, is a conference that illustrates the difficulties in getting where the title, *Beyond Whistleblowing*, points.

researchers have called attention to wrongdoing in research on human and animal subjects.[14]

For the whistleblower the problem is to gain a responsive hearing, while at the same time protecting his employment and professional standing. For the institution the problem is to remain receptive to internal dissenting opinions, while maintaining stability and guarding against abuse by self-aggrandizers and cranks. Usually the expectation by all parties is that a dissenting professional will protest internally first, escalate up through the organization as necessary, and appeal to outsiders (professional organizations, legislatures, the general public) only as an extreme resort.

Many institutions now have established administrative provisions for hearing internal dissent, getting the facts straight, protecting whistleblowers from reprisal, providing higher-level and possibly outside review, resolving complaints, and reinstating mavericks within the organization. A number of federal agencies have developed policies and procedures for protecting employees and facilitating "expression of differing professional opinion."[15]

Some technical societies, such as the Institute of Electrical and Electronics Engineers, have developed means for supporting their members in taking stands on unconscionable situations. For most such organizations this includes establishing an ethics committee, developing a code of ethics, and instituting ways for reviewing complaints. In a few cases societies have contributed financially to the legal defense of a whistleblower. But in 1980 the Professional Ethics Project of the American Association for the Advancement of Science had to conclude (Chalk, Frankel, and Chafer, 1980, p. 103):

> The availability and use of sanctions and supports are not well developed areas of [technical] society activity. Where sanctions do exist they are used infrequently. And when sanctions are applied to members, the societies generally do not inform other members or non-members of the action taken. Members who seriously seek to comply with their society's ethical rules can expect very little in the form of support activities.

RES PUBLICA

"The people is a myth, an abstraction," Carl Sandburg declaimed in "The People, Yes." So are "society" and "the public." Although I can't develop

---

14. U.S. President's Commission for the Study of Ethical Problems in Medicine and Biomedical and Behavioral Research et al., 1982.

15. Chalk and von Hippel, 1979, recommends some due-process procedures and surveys relevant federal provisions.

a theory of society here, I must at least survey some problems of civil order and *res publica* ("things public") that form part of the milieu of technical work.

"Public Goods" and "The Public Interest"

Generally we know what is meant by "public goods": resource and environmental goods, such as air and water, that are held and used in common by many people; or programs, such as hoof-and-mouth disease inspections, that serve regional or other aggregate interests. But how should commonwealth radiofrequencies or waterfowl breeding grounds, in which there can be many mutually conflicting ownerships and interests, be treated? How should policies and practices, such as quarantines and vaccinations, be cast so as to achieve individuals' submission to such collective societal goals as health protection? How should airspace, or quiet, or access to beaches, whose value is not readily accounted in economic markets, be controlled?

Technical people often become involved in debates over whether goods should be treated as public goods politically. For example, biologists have helped gain public-goods status for ecologically important but underappreciated marshlands. Many social scientists have protested against the way the 1980 U.S. Census, a public good provided for by the Constitution, became commercialized by the Reagan administration (the results were released first on computer tape, rights to purchase the computer data according to postal Zip Code were sold to a consortium of data-service firms for resale principally for market research, and the census was printed by the government only slowly and in limited form) (Starr and Corson, 1985).

Related is the problem of "the public interest." A public, as *Britannica* defines it, is a group whose members attend to a common set of issues, interact with each other at least indirectly, and are aware of their commonality of interest (*Macropedia* 16, p. 960, 1974). Walter Lippmann argued that "The public interest may be presumed to be what men would choose if they saw clearly, thought rationally, and acted disinterestedly and benevolently." " . . . and understood the technical complexities," we would have to add.[16]

But the fact that there are countless publics, having countless interests, debilitates many "public interest" arguments and "public participation" claims. Ironically, matters alleged to be in the public interest often don't interest the general public very much.

In the past two decades many groups have taken up "public interest" and "environmental interest" and "consumer interest" banners. Thus

16. Lippmann, 1955, p. 42. For a recent discussion of public interest and civic rhetoric, see Stanley, 1984.

there are public interest research groups, public interest law firms, and organizations like Medicine in the Public Interest (hard to describe, but its name implies that other medicine is pursued against some public interest). The self-enfranchisement in such institutional personas is becoming more evident, now that both pro- and anti-nuclear-power organizations are calling themselves public interest groups. Traditionally such organizations have thought of themselves as speaking for causes that can't speak for themselves (endangered biota, wilderness areas), or that are poorly organized (some workers, consumers), or that otherwise are not fully staked in established political arenas or markets.

Regardless of whether I agree with them on substance, I am annoyed when groups I do not belong to bluff and, claiming to "represent the public interest," take a stand on which I, a member of the public, neither have been consulted nor have delegated my representation. Because particular publics and interests must be defined for any case, it is more authentic to call such groups "special interest groups," and to avoid referring to "public interest" unless one can specify both the *public* and the *interest* in the case and confirm that the public being spoken for agrees to, or clearly can be expected to agree to, the representation.

Society being a "set of sets" (Fernand Braudel, 1979, p. 458) or a "social union of social unions" (John Rawls, 1971, p. 527), in any instance one can define what set is being referred to, or allow that the boundaries are ill-defined or permeable. The term "society" remains useful. Thus I don't think it confusing to say, as I have many times in this book, that society decides something or takes some action. The abstractness and flexibility of the notion don't disable it. During any social movement it is, in fact, important to ask: In what sense are we/they acting *as* society?

Political theorists continue to redefine democracy, the state, representation, public welfare, and related abstractions. Many scholars construe the state, and political officeholders, principally as brokers of competing interests. Charles Lindblom, adopting Robert Dahl's term "polyarchy," has portrayed the United States and the nations of Western Europe as "containing polyarchy" (Dahl, 1956, p. 74; Lindblom, 1977, p. 131). E. E. Schattschneider, believing that "the socialization of conflict is the essential democratic process," has argued that "democracy is a competitive political system in which competing leaders and organizations define the alternatives of public policy in such a way that the public can participate in the decision-making process" (1960, p. 141).

Public Participation

But the notion of public participation in decisionmaking also is problematic. No one person or group of people fully represents, or is representative of, the citizenry, and the organized or attentive public usually is small and keeps changing in composition and opinion.[17] Openness in bureaucratic proceedings usually is desirable, but it is difficult to devise proceedings that encourage public participation without merely opening up to narrow-interest lobbying. Policymaking may or may not be enlightened by the opinions of a few individuals or pressure groups.

Public, or lay, participants indeed may contribute to deliberations. More often than not these participants are citizens distinguished in community or professional affairs. If they play their role seriously they may canvass and express prevailing lay opinion beyond their own. They may bring up factors beyond the range raised by the experts, recruit a diversity of other voices into the discussion, and smooth the group's relations with the general public. Lay participants can be extremely valuable to human-subjects research review boards, national commissions, advisory groups of many kinds, and some R&D funding committees. But appointing a few citizens distinguished mainly for their miscellaneity (such as, in bygone days, "one intellectually divergent redheaded housewife") to boards doesn't guarantee more astute or representative deliberations.[18]

CONFLICT BETWEEN EXPERT AND LAY OPINION

In an op-ed letter in the *New York Times* (January 31, 1981) biochemist Rosalyn Yalow complained about the closing of sites for disposal of low-level radioactive waste from biomedical research: "The Nuclear Regulatory Commission should have proposed a change in regulations long before the sites were closed down. Why didn't it? The regulators were responding not to the *real* risk but rather to the public's fears of radiation at any level. . . . We must respond to real risk, not perception of risk."

Disputes like this one strongly test republican forms of government, in which legitimate authorities, in government and elsewhere, are charged with making determinations on behalf of the public. This is one of the central problems of our times.

The clash may reflect conflict either between some version of technical rationality and other opinion, or between differing value stances held by the protagonists. Rarely are all experts lined up on one side and all nonexperts on the other. Usually, as more people become engaged in a controversy, one growing expert-and-lay coalition becomes aligned against another growing expert-and-lay coalition.

As I have argued all along: although all thinking is by definition subjective, questions such as what relative contribution different sources make to people's radiation exposure can, by scrutiny within the organized skepticism of science, become authenticated into reliable knowledge, knowl-

17. Fishkin, 1981, is a symposium on the theory and practice of representation.
18. Organisation for Economic Cooperation and Development, 1979.

edge that describes and predicts with consistency and accuracy *what is* and *what will happen* in the physical and biological world. While I might evaluate Dr. Yalow's radwaste differently from the way she does, I too protest that I don't want my radiation protection determined by public opinion polls! But I hasten to say that I object to the arrogant term "real risk"; one person's objectivity is another's subjectivity, and, in science as elsewhere, countless factors influence definition of the real. What has to be sorted out are differences between "the real" as intuited and "the real" as constructed by expert consensus and tested against experience.

Just as troublesome a view is the sort economist Lester Lave expressed to a conference on dioxin (1983, p. 636):

> If the public generally believes that a death from cancer is many times worse than death from other causes, regulations must embody this value. If the general public, or an articulate group claiming to represent them, believe that accidents are many times more likely than scientists estimate, stonewalling will be dysfunctional.

No, stonewalling won't help; but comparative evaluation as described in Chapters Seven and Eight can foster stable change and reasonable allocation of resources. When the facts about an issue become clear, and consensus on values solidifies, the public can establish any regulations it wants. But the polity needn't act on mere intuitions, and certainly not on the opinion of any "articulate group claiming to represent" the public's interest. We must resist letting these complex controversies become too simple.

Some extreme-subjectivist social scientists and journalists perpetrate the view that since scientists sometimes abuse fact/value distinctions and often disagree among themselves over the facts, and because the problems are too complicated to be factored apart empirically, only people's perceptions matter. In this vein an essay in the *New York Times* once headlined, "In the Human Equation, Risk Perceived is Risk Endured" (Browne, 1980). Would that this were so. To the contrary, alas: Risks *Not* Perceived May Well Be the Ones that Get Us. Likewise, benefits perceived are not necessarily benefits gained, and potential benefits overlooked may be benefits forgone.

Efforts now are being made to study how people develop their perceptions of such matters as health or environmental risks and benefits. People's attitudes on health risks seem to be influenced by such factors as voluntariness of exposure, frequency and familiarity of occurrence, amenability to personal control, reversibility, immediacy of effect, equitability, bizarreness, and catastrophic potential.[19]

Social scientists are applying opinionnaire techniques to surveying peo-

19. Lowrance, 1976, pp. 86–94. For an attempt to develop a "multivariate conceptualization of hazardousness," see Hohenemser, 1983, pp. 49–67.

ple's perceptions and their risktaking and costbearing proclivities. What they find, to neither their surprise nor ours, is that people have perceptual biases. This research has concluded that human beings' brains, whether expert or lay, become overloaded with information and have trouble making comparisons; that the newsmedia accentuate social reverberations in disputes; and that people believe what they want to believe. Most of these polling studies are open to criticism. They suffer from the usual shortcomings of questionnaire design and the generic weaknesses of polling. Often they ask about only a single issue at a time, which by failing to foster or force comparison, and by allowing people to express self-contradictory views, provides little guidance for policymaking. One of my main reservations stems from the fact that perceptions are subject to rapid shifts: hydropower, which generates "clean" electricity and incidentally creates enjoyable large lakes, probably will continue to be viewed as harmless, until the day a large dam breaks over a major population center.[20]

Several theories of perception and judgment under uncertainty are now being developed that are helping make sense of how various observers— nonexpert or expert—come to hold the views they do. Eventually these theories may suggest ways for going about reconciling differing opinions.[21]

Rightly concerned about the polity's reception of complex probabilistic information, Baruch Fischhoff posed a core problem (1977, p. 189):

> At some time in the not too distant future, those of us living in earthquake-prone areas may receive messages like the following: "There is a 50% chance of an earthquake of magnitude between 6.5 and 7.5 along a fault line of 10–50 miles centered approximately 50–100 miles south of town to occur 3 years from now, plus or minus 6 months." Will we know how to respond to the gamble this message implies?

No, not by ordinary commonsense analysis. Some inhabitants will panic and evacuate immediately, taking little damage-limiting action, while others will stay to "ride it out" and protect their property; and later both groups will demand public compensation, either for quake damages or for protective expenses or losses of opportunity. The scientific messengers will be asked, "What damage will the earthquake do? And exactly when will the earthquake hit?" Modern society's commitment to collective, mutually assistive risk buffering calls for expert critiquing of such forecasts, expert interpretation of the options and expected consequences, expert comparison of the situation to other such gambles, and expert help in structuring

20. Fischhoff et al., 1981, summarizes some of these studies and their use in decisionmaking. Cole and Withey, 1981, is a solid review. A polemical essay is Royal Society Study Group on Risk, 1983, pp. 94–148. Covello et al., 1983, doesn't make much progress but reveals the shape of the issues.

21. Kahneman, Slovic, and Tversky, 1982, collects stimulating articles.

the facts and values into decisions. Only with such analysis can it become clear which gas and electric lines to shut off, which hospitals to close, which neighborhoods to evacuate, which highway bridges to barricade, which disaster procedures and compensation schemes to adopt. "Generalship" will be crucial. The development from alarm to plan will have to be iterative and integrative: public concern, expert response, public reaction, expert reiteration, public plan, expert critique, revised public plan, public program. . . .

Frustration with such affairs may lead to misgivings of the kind Steven Rhoads expressed in a 1978 article on an analogous issue, "How Much Should We Spend to Save a Life?" (1978, p. 92):

> As for candor, I would tolerate a little dissembling in this area. . . . Let analysts and political administrators argue that the willingness-to-pay figures and other explicit values are just one factor in the decision-making process. Let interested Congressmen have the analyses so that they can better appreciate the policy dilemmas, but let them also sympathize with their constituents' anger and pledge to look into the matter. If Congressional hearings are held, one can be sure that they will not lead to a legislated value for life—and that is all to the good. Admittedly, the absence of total candor leaves the way open for the periodic appearance of ambitious reporters and politicians who expose the Dr. Strangelove-like analysts at the heart of the bureaucracy. But this cost seems tolerable in an area where simple enlightenment is not to be expected or wished for.

Reminiscent of the thrush's remark in T. S. Eliot's "Burnt Norton" that "Human kind/ Cannot bear very much reality," Rhoads's reservations are understandable, though patronizing. Candor often is painful. But "a little dissembling" is no solution. Anyway, these tragic choices now are being discussed openly—to considerable long-term benefit, I am convinced, despite the discomfort they cause.

From now on, the technical complexities of defense, telecommunications, agricultural development, environmental health, and other such issues simply will not be comprehensible in depth to the lay public. Outside of their specialties scientists will not find it easy to evaluate these problems, either. Experts from different fields will have to struggle to assemble the "big picture." What we will need will be mechanisms for sorting out differing expert opinions, and sensitive procedures for integrating technical aspects with broader societal aspects.

MECHANISMS FOR MANAGING SOCIOTECHNICAL DISPUTES

Most proposals for "managing" sociotechnical disputes are attempts to accommodate the adversarial nature of disputes while forestalling win-at-all-costs game-playing and acrimony, and to delineate controversy and demarcate consensus. They are meant not to stop arguments, but to generate findings that carry legitimacy and make contributions to the long-term agenda. Mechanisms that are being proposed and tried include new forms of government hearings, public inquiries, versions of "science courts," consensus-development conferences, and environmental mediation.

Government Hearings

Many federal agencies have upgraded their hearings procedures. In an effort to develop full dossiers on the pesticides DDT, dieldrin and aldrin, heptachlor, and other chemicals, the U.S. Environmental Protection Agency (EPA) has asked administrative law judges, appointed by the Justice Department, to conduct lengthy hearings on many aspects of the pesticides. Interested parties file statements, and the judges may ask for additional materials. The parties do not debate each other directly. The resulting record, with the examiner's conclusions, is provided as advice to the EPA administrator.

To explore the issues surrounding the building of a large-diameter, high-pressure gas pipeline from the new Prudhoe Bay and Mackenzie Delta oil fields down to the United States, in the mid-1970s Canadian Justice T. R. Berger conducted an elaborate, well-regarded Mackenzie Valley Pipeline Inquiry into energy policy, environmental hazards, industrialization strategy, and cultural implications of the project. Justice Berger held four different sets of public hearings: Formal Hearings to receive evidence on engineering proposals and expected environmental impacts; Special Hearings on the gas companies' exploration and production programs; Southern Hearings to elicit citizens' views in the large cities in the lower half of Canada; and Community Hearings in the directly affected northern areas. The report, *Northern Frontier—Northern Homeland*, became a national best-seller.[22]

In 1977 the U.K. Department of the Environment conducted a wide-ranging, quasi-legal inquiry into a proposal by British National Fuels Limited to build a billion-dollar plant at Windscale, West Cumbria, to reprocess spent British reactor (thermal oxide) fuel and possibly foreign nuclear fuels as well. The local inquiry quickly escalated into a reexamination of national nuclear energy policies. Most parties were represented by legal counsel throughout. Commentaries on the inquiry have revealed major dissatisfactions: that the burden of proof was laid on those who objected to the proposal; that the scope of the hearings was not clarified firmly at the outset; that the inquiry was too legalistic in character; that not enough time or attention was devoted to preinquiry assessments; that little provi-

---

22. Berger, 1977, is the report of the inquiry. Gamble, 1978, is a review.

sion was made for funding of intervenors; and that the hearings were conducted as a stream of for-and-against testimonies, not organized by substantive issue.[23]

"If the Windscale Inquiry can be partially characterized as a trial between competing facts and competing logics," a review of these two hearings observed, "then the Mackenzie Valley Pipeline Inquiry was a consciousness-raising public teach-in." It concluded (Organisation for Economic Co-operation and Development, 1979, pp. 66–77):

> Inquiries that proceed by a structural discussion of each major issue, as opposed to some more arbitrary sequence of proponent and opposition intervenor groups, not only have the advantage of reducing the presentation of repetitive testimony, but, more importantly, of developing a cumulative information base from which to assess cumulative impacts. Recourse to the adversarial approach to the examination and cross-examination of witnesses and testimony can often result in more critical information becoming available, in the more thorough and penetrating examination of competing claims, and in the clearer articulation of individual and groups' interests and biases. . . . It also has its disadvantages, however, especially when it results in sometimes exclusive reliance on lawyers and technical experts to the detriment of direct citizen participation.

Such U.S. agencies as the Consumer Product Safety Commission and the Nuclear Regulatory Commission have conducted public hearings on sociotechnical matters. Most of these have been passive affairs not allowing much direct confrontation or debate, and seem to have served mostly as means for sensing the tenor of concern and mollifying some agitated publics.

"Science Court" Proposals

A remedy often proposed is a "science court." Described in the early 1960s by Arthur Kantrowitz and promoted by him ever since, the unfortunately named "court" has provoked discussion but has never, so far as I know, actually been tried on an important problem. As Kantrowitz describes his court:[24]

> The basic mechanism . . . is an adversary hearing, open to the public, governed by a disinterested referee, in which expert proponents of the opposing scientific positions argue their cases before a panel of scientist/judges. The judges themselves will be established experts in

23. Parker, 1978, is the record of the inquiry. Breach, 1978, and Wynne, 1982, are commentaries.

24. Kantrowitz, 1977; earlier versions are Kantrowitz, 1967, 1975.

areas adjacent to the dispute. They will not be drawn from researchers working in the area of dispute, nor will they include anyone with an organizational affiliation or personal bias that would clearly predispose him or her toward one side or the other. After the evidence has been presented, questioned, and defended, the panel of judges will prepare a report on the dispute, noting points on which the advocates agree and reaching judgments on disputed statements of fact. They may also suggest specific research projects to clarify points that remain unsettled.

In 1976–1977 a task force appointed by the White House explored the possibilities of a science court experiment, and produced an interim report recommending that the tribunal be tested on some "fairly simple issue" such as food dyes or water fluoridation.[25] Despite willingness by at least one regulatory agency to refer an issue to such a court, the experiment was not attempted. Many difficulties would remain in defining the issues, selecting participants, dealing with fact/value confusions, and so on, but in principle most of these could be worked out.[26]

However, it seems to me that a restricted version of science court that dealt with delineated, routine technical disagreements would be no different from well constituted agency staffs and advisory committees that already perform admirably, or have potential to. And an ambitious version of the court, such as a tribunal addressing (to use one of Kantrowitz's examples) whether a particular nuclear power plant should be licensed, would find itself usurping the decisional prerogatives of the agency its findings would be directed to. Unless chartered by the Nuclear Regulatory Commission to conduct reactor licensing hearings under the guidelines and accountability of federal law, how could it possibly perform that function? But if chartered specifically for that purpose, how would it be different from the Commission's own duly conducted licensing hearings? The court wouldn't, in my opinion, fill any new niche. But the spirit of the remedies the proposal embodies should be instilled wherever possible into existing mechanisms.

Central Authority Structures

For appeal of technical disputes, central authority structures may be proposed. An extreme, and I think unworkable, example was Simon Ramo's recommendation to separate the duties of investigating the "scientific disbenefits" from those of "balancing the good against the bad aspects of

25. Task Force of the Presidential Advisory Group on Anticipated Advances in Science and Technology, 1976; and Boffey, 1976.

26. For commentary, see Casper, 1976; James Martin, 1977; Mazur, 1977; American Bar Association, 1979; and Michalos, 1980.

various alternatives and making decisions" in regulation (1981). Ramo would have the former conducted by a single huge federal scientific agency, which would report its findings to several executive branch "decision boards" appointed by the President. Each board would then "have the role of comparing alternatives, balancing the good against the bad, and the duty to connect the case before it to other national interests. It should have the unquestioned responsibility for banning or approving the challenged technological operation." Since the very planning and funding of scientific assessments is inherently social and political, such a superagency could never be "purely scientific." Just deciding which technical questions to place on the agenda is value-laden. Besides, there is no way such a huge laboratory could tailor its scientific investigations to the myriad highly specialized needs of boards having the regulatory scope of current agencies. Ramo's decision boards otherwise appear to be equivalent in mandate with those of agencies already in place.

Proposals intending to institutionalize the separation of science from other aspects of judgment have continued to arise, as this one did in 1981 (American Industrial Health Council, 1981):

> The American Industrial Health Council advocates that in the development of carcinogen and other federal chronic health control policies scientific determinations should be made separate from regulatory considerations and that such determinations, assessing the most probable human risk, should be made by the best scientists available following a review of all relevant data. These determinations should be made by a Panel of eminent scientists located centrally somewhere within government or elsewhere as appropriate but separate from the regulatory agencies whose actions would be affected by the determinations.

Two questions must be asked of such proposals: whether "scientific and technical determinations" can legitimately be separated from "political and social determinations," and whether centralization of authority assures higher-quality science. It should be clear by now that to the first I would answer, Yes, to a considerable extent, as long as it is understood that the very process of defining problems is judgmental, that meta-analytic issues may be pivotal, and that scientific assessments usually have to be conducted iteratively. Complex issues, such as energy policy, should be expected to go many rounds of assessment, criticism, redefinition, and reassessment. To the second question I would answer that communal scientific assessments do tend to gain critical analytic strength and social legitimacy over assessments made by individuals. But pluralism is an essential safeguard against narrowness; centralization and consistency in themselves guarantee nothing. Pluralistic diversity should be fostered by recruiting skilled policy-sensitive scientists and engineers to industry, government, and other organizations, by appointing able advisory panels to many different administrative, legislative, and managerial bodies, and by upgrading assessment work in academies, professional societies, and labor and industry organizations.

Given the hundreds, thousands, indeed tens of thousands of issues that actually reach decisionmaking or lawsuit levels, and the many more that have the potential to, merely establishing blue-ribbon advisory panels may help but will not solve the problem. It would only take a few issues having the national importance of asbestos, diesel emissions, swine flu, acid rain, or 2,4,5-T to saturate the committee's capacity. If a national regulatory science panel is established, its efforts should be reserved for rendering long-term guidance of legislative, budgetary, and program development, and for critiquing generic principles used by the hundreds of staff and advisory groups that unavoidably must, with meticulousness and doggedness, conduct assessments week in and week out. Only very occasionally should the panel take upon itself the task of delving into particular assessments in detail. In 1983 a National Research Council committee examined the American Industrial Health Council proposal quoted above and recommended that "regulatory agencies take steps to establish and maintain a clear conceptual distinction between assessment of risks and consideration of risk management alternatives," that "uniform inference guidelines be developed," and that "a Board on Risk Assessment Methods be established."[27]

Other Aspects of Controversy Resolution

We must continue to seek ways for resolving issues without having to go through polarized, acrimonious, socially disruptive battles. I hope some of the attitudinal and procedural recommendations of this book will help. For the most part I think these requirements can be met by current institutions. There is a need, though, for imaginative and courageous reform within those institutions. And some new institutions are showing promise.

Often, of course, controversies partly concern social authorities' management of affairs. Nowhere has this been more bluntly evident than in the primary conclusion of the Kemeny Commission's report:[28]

> To prevent nuclear accidents as serious as Three Mile Island, fundamental changes will be necessary in the organization, procedures, and practices and—above all—in the attitudes of the Nuclear Regulatory Commission and, to the extent that the institutions we investigated are typical, of the nuclear industry.

27. NRC, Committee on the Institutional Means for Assessment of Risks to Public Health, 1983, p. 7.

28. U.S. President's Commission on the Accident at Three Mile Island, 1979a, p. 7.

Public confidence in the decision process obviously is important, as is a sense that authorities are technically competent and acting in good faith. These factors are especially important in public disputes that are presented as centering on technical issues but that are proxy complaints about corporate bullying, or bureaucratic turgidity, or erosion of personal control.

Assessment and decisionmaking are made less disturbing to the public and more civil as process: if legislative mandates are made detailed and clear; if technical ground rules are critiqued firmly not only at the end but at the beginning of each major phase of assessment or appraisal; if agencies build reputations for technical and political integrity and legitimacy, including having advisory apparatuses in place, *before* emergencies arise; and if organizations devise ways for responsibly saying (by actions as well as words), when necessary, "We don't know."

A number of new institutional forums have been created to seek clarification and consensus on sociotechnical problems. The National Academy of Sciences and its affiliated organizations have upgraded their study efforts. The Food Safety Council, a coalition of food industries and nonindustrial groups, prepared a well-regarded report on scientific aspects of a *Proposed System for Food Safety Assessment*; less successfully, it also drafted a prospectus for a food safety regulatory procedure (Food Safety Council, 1980, 1982). In 1983 the National Center for Toxicological Research, along with a wide variety of co-convenors, held a scientific Consensus Workshop on Formaldehyde (*Environmental Health Perspectives*, 1984). My own program at The Rockefeller University has convened technical symposia on Assessment of Health Effects at Chemical Disposal Sites and on Public Health Risks of the Dioxins (Lowrance, 1981, 1984). The U.S. Food and Drug Administration delegated the scientific review of aspartame, the artificial sweetener, to a public board of inquiry; although the procedure has been criticized as faulty, lessons can be drawn from it.[29] In some localized environmental controversies, mediation is proving helpful (Allan Talbot, 1983). In 1983 a Dialogue Group on Hazardous Waste, under baroque auspices involving experts and citizens of diverse affiliations and views, prepared an excellent handbook of questions entitled *Siting Hazardous Waste Management Facilities*.[30]

Elements that can help damp the reverberations of disputes include:

- Improving quality control of the science used in public decisionmaking;
- Building foundations of decision-relevant recordkeeping (such as improved health statistics) and research (such as evaluating the predictive validity of environmental assays);

29. Brannigan, 1983, is a commentary.

30. Available from the National Audubon Society, 115 Indian Mound Trail, Tavernier, Florida 33070.

- Ensuring that all aspects of assessments are made explicit and on record, that uncertainties are described, and that social value judgments are identified as such;
- Subjecting major sociotechnical assessments to broad critique;
- Conducting arguments in an open, good-faith atmosphere, and providing for public participation wherever appropriate;
- Drawing lessons from landmark cases;
- Instituting means for structuring agendas so as to minimize surprises and to foster stable consideration of issues; and
- Developing quasi-governmental and nongovernmental organizations that can contribute to evaluation and dispute resolution.

At least as important as these procedures is the "filling of the middle" in controversies by experts who do not represent extreme technical or social views. They can raise questions, bring their expertise to bear on the issues, and help foster civil and technically sound discussion. Along with this we need compensation against the perverse effect that, as a sociotechnical controversy gets messier and more political, as the issues become distorted, as factual authority breaks down under uncertainty and disingenuous attack, competent and well-intentioned experts may back away from involvement—at the very stage where their participation is needed most—out of disgust at the fracas. We should hope that technical organizations will provide supportive auspices for the rendering of expertise in controversies, both in providing forums and in legitimizing and reinforcing individuals who are willing to try to make a contribution.

STRUCTURES FOR STRUCTURING AGENDAS

Our enormous current toxic waste problems unquestionably could have been ameliorated if they had been given R&D and managerial attention earlier. For too long we neglected municipal wastes, chemical manufacturing wastes, and radioactive wastes. Belatedly, now, at substantial agony and expense, we are working out methods for reducing waste, reclaiming materials from waste, containing and detecting leaks from storage basins, and incinerating materials efficiently. All of these were technically within reach years ago. The only reasons given why that reach was not attempted are banal: the problems grew by quiet, slow accumulation, and they were not viewed as being technically challenging or managerially important. As Carroll Wilson, the first general manager of the Atomic Energy Commission, recalled of radwaste: "Chemists and chemical engineers were not interested in dealing with waste. It was not glamorous; there were no careers; it was messy; nobody got brownie points for caring about nuclear waste. The Atomic Energy Commission neglected the problem" (1979, p. 15). The same could be said of the other waste problems.

Improvement of our mechanisms for setting sociotechnical agendas would reduce disruptive surprises and anxieties, foster anticipation of problems, help organizations plan their allocation of resources and attention, and lead to more orderly development and resolution of controversies.

Scientists can take initiative. Since 1975 chemists from forty-three nations, working under the aegis of the International Union of Pure and Applied Chemistry, in a program called Chemical Research Applied to World Needs, have held major conferences on future nonpetroleum sources of organic raw materials, such as biomass and untapped fossil deposits, and on the application of chemistry to feeding the eight billion people—twice the number of 1975—expected at the world's dinner tables by the year 2015; their purpose has been to influence governments' indigenous R&D agendas.[31] In a 1984 study from the Worldwatch Institute, William Chandler proposed a least-cost strategy for improving world health, based on "providing maternal and child care for the world's poorest people, clean drinking water and sanitation facilities to the third of the world's population that lacks them, diet education for populations at high risk of heart disease and cancer, control of tobacco products, and basic research for low-cost cures" (1984, p. 7).

Overall the agenda-setting problem has two aspects. The first is how to articulate uncertain estimates and tentative concerns into inducements or warnings, without either making unwarranted promises or alarming the public unduly. The second is how to convert the inducements or warnings into *agendas*—literally, "actions that must be carried out."

Take the problem of soil erosion. A 1981 Department of Agriculture national survey estimated that the United States is losing about two billion tons of cropland soil every year. Two billion tons, irrecoverably. The problem is diffuse and is manifested very differently in different regions. In some places, in the not-too-distant future, erosion could be as bad as that incurred during the dustbowl era. How should attention be drawn to the problem? How should the R&D agenda be set? How should remedies be decided on and implemented?[32]

Or take groundwater contamination. Hydrological experts have expressed concern for years. Yet nobody is squarely "in charge" of the nation's groundwater. Many public agencies' authorities overlap, many private interests are involved, and international diplomatic relations are at stake. Steadily American aquifers are becoming polluted as wastes percolate into them. Again, how to install this problem firmly on the nation's action agenda? How to present it *as an issue*? How to coordinate the undertaking of reliable analyses and surveys? How, then, to organize the

---

31. St.-Pierre, 1978; Bixler and Shemilt, 1983; and Shemilt, 1983.
32. Batie, 1983. For historical background, see Hyams, 1976.

inevitable debates? How, eventually, to translate the emerging facts and value judgments into, say, a national aquifer protection plan?[33] The same wrenching questions can be asked about global climate change, destruction of upper atmospheric ozone, tropical deforestation, the developing countries' population explosions, and many other such issues.

Major assessory studies can help. So can "prioritizing exercises." A committee of the Institute of Medicine recently proposed a system for establishing priorities for new vaccine development. The method is based on assigning, for each vaccine, numerical answers to the commonsensical questions: How many people are made ill by the disease every year? How severe is the illness? How much in medical care costs would the vaccine save? What groups are most susceptible? Is enough known about the disease organism to develop a vaccine? How much will vaccine development cost, and how long will it take? For whom should vaccination be recommended? How much will a vaccination program cost?[34]

Some observers have called for Councils of Urgent Studies (Cellarius and Platt, 1972). Such organizations as the Worldwatch Institute, Resources for the Future, and the national science academies provide analysis-and-warning, and such new organizations as the World Resources Institute intend to. President Carter established a Commission for a National Agenda for the Eighties to nominate problems for attention; but, published at the end of that administration, the report seems to have been ignored by the Reagan administration and the Congress. Veterans of public service often recommend the establishment of permanent bodies to broker the setting of agendas on complex matters.[35] Robert Keohane and Joseph Nye proposed instating continual "long-term scanning" of global environmental and resource issues, with a view toward anticipating problems, identifying policy and R&D opportunities, and stimulating responses from bureaus (1975). This function is important. High on our agenda should be the exploration of ways for securing it.

33. Pye, Patrick, and Quarles, 1983; American Institute of Professional Geologists, 1983; National Research Council, Panel on Groundwater Contamination, 1984; U.S. Environmental Protection Agency, 1984; U.S. Congress, Office of Technology Assessment, 1984; and Gordon, 1984.

34. Institute of Medicine, Committee on Issues and Priorities for New Vaccine Development, 1984.

35. Cahn and Cahn, 1980, reviewing the experience of high-level commissions 1908–1978, found that most of the task forces recommended the establishment of permanent agenda-setting offices.

Ten

Stewardship Beyond
Narrow Responsibility

After probing the possibilities and limitations of responsibility in Chapter Four, I concluded:

> Wherever contractual or traditional expectations are clearly agreed on, responsible action by practitioners obviously should be expected and enforced. But, especially for important complex problems, non-specific responsibility can neither sufficiently motivate nor provide guidance for individual scientists' actions. Even extraordinary efforts by extraordinary individuals will be inadequate against some of the novel and diffuse problems we face today. Collective approaches must be sought that do not depend on heroic initiative or sacrifice. These approaches must be based not so much on legalistic compulsion as on inducement and stewardship.

"Stewardship" is the term that conveys the necessary bivalent meaning: nourishing science for internal intellectual ends, and, at the same time, orienting and applying the science to external socially valued ends.

At present, even for such well-defined areas as medicine, many philosophers and critics are wary of the notion of collective responsibility. Peter French concluded that duty lies both with "the profession" and its members (1982, p. 84; emphasis added):

> The medical profession *is* responsible for the inequitable delivery of health care in the United States. That does not mean that the profession is solely responsible, but . . . that the individual members of the

profession bear the moral burden for the problem. . . . The use of the name of the aggregate in such a responsibility ascription puts each and every one of them "on call."

There is every reason, I am convinced, to construe today's "profession and practice of medicine" as being larger than the sum of the services of individual doctors to their individual clients. We must hope that physicians will volunteer, and *put each other on call*: to evaluate practices, to improve the education and licensing of new colleagues, to anticipate and address new health issues, to revise the social context of the profession (including access to health care), to prescribe reforms. This is a consequence of their profession's compact. Reformers like Abraham Flexner will come along from time to time, and we must encourage them, but we can't wait for their advent. The traditional view that physicians owe respect only to their particular patients, and for clinical care alone, is, to me, simply and obviously inadequate.

In 1984 the Judicial Council of the American Medical Association ruled: "To expect a physician in the context of his medical practice to administer governmental priorities in the allocation of scarce health resources is to create a conflict with the physician's primary responsibility to his patients that would be socially undesirable" (American Medical Association, 1984, p. 3). In his proposed medical covenant, ethicist Robert Veatch also defined the doctor's world narrowly, stipulating that: "Individual practitioners shall be exempt from the general moral requirements of the principles of justice, including their impact on health care planning and cost containment, insofar as they are committed to patients in ongoing lay–professional relationships" (1981, p. 330). Some medical ethicists take a broader view. In an essay, "The Hippocratic Ethic Revisited," Edmund Pellegrino emphasized the collective duty (1979, p. 112):

> One of the gravest and most easily visible social inequities today is the maldistribution of medical services among our population. This is [a] sphere in which the profession as a whole must assume responsibility for what individual physicians cannot do alone. . . . We must engender a feeling that the entire profession suffers from ethical diminution whenever segments of the population lack adequate and accessible medical care.

This debate will have deep consequences for the future of medicine and all of health care. Movement is occurring. In 1980 the California Medical Association brought legal suit asking a federal court to compel Governor Brown's administration to provide better acute medical care for the mental patients at Lanterman State Hospital and other "grossly substandard" facilities (the judge dismissed the case on the ground that the physicians were not representing particular patients).[1] In 1984 a task force of the American Psychiatric Association prepared a thorough study, *The Homeless Mentally Ill*, in the recognition that "while all citizens have a responsibility for the welfare of the homeless, psychiatrists have an additional responsibility for the mentally ill among them" (American Psychiatric Association, 1984).

Engineering too has tended to define itself atomistically. An otherwise illuminating review of the legal and moral responsibilities of engineers set the issue narrowly in its first sentence: "The engineering profession is composed of individual engineers, so that engineering will be responsible if its individual member engineers are responsible" (Mingle and Reagan, 1980, p. 15). The authors, themselves engineers, described the legal liabilities to which engineers are subject. Then they pointed out how, in incidents like the first DC-10 disaster and the BART brakes episode, the demands on contemporary engineering hardly can be met by individual responsibility. But they failed to go on and conclude, as I did of physicians: The profession and practice of engineering are much larger than merely the sum of the acts of individual engineers. How can any individual engineer, on his own, know "his part" in the profession? Who, other than engineers-in-collective, is going to reform engineering education, or enforce standards of practice, or advise public authorities on matters of engineering? The very powers of engineering derive, in fact, from its standardizations and its systematized exchanges of lessons from experience. These should be built upon.

The kind of compact-based stewardship we need in many areas was exemplified by recent debate in a much less formally professionalized discipline. Concerned about the erosion of federal statistics programs, James T. Bonnen of Michigan State University testified to a committee of the House of Representatives:[2]

> The protection of the integrity of statistics has its foundation in the integrity and courage of the statisticians, demographers, economists and other analysts who design and produce statistics. Since isolation from the policy process is neither desirable nor possible, the institutional safeguards to integrity should involve appropriate processing and publication standards, publication of methods, a well articulated legislative mandate for individual statistical agencies, a strong common confidentiality statute for all major agencies, high visibility and multiple accountability of statistical policy, a central coordination unit

1. *California Medical Association* v. *Brown*, no. C 79-3323 WAI, slip opinion (N.D. Cal., February 28, 1980); and Cooney, 1979.

2. U.S. House of Representatives, Legislation and National Security Subcommittee, of the Committee on Government Operations, 1982, p. 391; statement also published as Bonnen, 1983.

for statistical policy with statutory responsibility including the integrity of federal statistics, and a single committee in each house of Congress for legislation and oversight of multiple purpose statistics and government-wide statistical policy and priorities. . . . Polite, reasoned conversation among the converted will no longer do.

Notice the salient features: an important technical resource, serving both basic research and applied societal uses; the resource nourished, tapped, and promoted by expert technical stewards of various affiliations, operating within a compact with society; and the scope, quality, and accountability of the programs supervised by diverse intersecting institutions. Most statisticians and social researchers seemed to agree with Bonnen Steps now are being taken to institute the required mechanisms.

WATCHING THE WATCHERS

Central among the questions of stewardship is the one so often cited in Juvenal's formulation, *"Quis custodiet ipsos custodes?"* "Who will watch the watchers themselves?"

The principal answers, I have suggested, are generated out of the grounding technical compacts: expert communities nominate guardians and stewards, such as advisory task forces; the polity scrutinizes the results, and occasionally presses for change; and, just as soldiers standing guard in Juvenal's day were expected to, the watchers watch each other.

Especially acute scrutiny can be exerted by technical communities whose interests flank or intersect those of the central discipline on an issue, the way microbiologists did on recombinant DNA. Cross-illumination can be very constructive. Epidemiologists, who retrospectively seek inferences about illnesses possibly associated with chemical exposure, complement the work of toxicologists, who experimentally test chemicals' effects on animals and people.

Over recent years the guardian functions of such established institutions as the National Academy complex, the Congress, and the federal agencies have been expanding, as I have described. Many nongovernmental organizations—the Conservation Foundation, the Environmental Mutagen Society, the Natural Resources Defense Council—have come on watch. A few groups, such as the Federation of American Scientists, long have contributed to sociotechnical debates. Many new "critical" groups, such as the Environmental Defense Fund and the European Environmental Bureau, gained impetus during the security and environmental controversies of the 1960s and 1970s. Such organizations as the British Society for Social Responsibility in Science, a socialist group founded in 1968, act as standing critics of the establishment; the Society is, in the words of its publicity material, "committed to fighting for the use of science and technology by and for the benefit of working people, to demonstrating the political content of existing science, and to furthering links between scientific workers and the rest of the labour movement." Some organizations, such as the [British] Council on Science and Society, involve people of diverse backgrounds and talents in working toward consensus reports (in the case of the Council, on such issues as nonlethal weapons for control of civil disorders). Others bring tough advocacy criticism to bear on government actions, as the Union of Concerned Scientists has in opposing the Reagan administration's "Star Wars" space-based missile defense proposal (Union of Concerned Scientists, 1984).

All such work needs to be evaluated on its own merits. No group is always right, and I don't mean to endorse any group here. Pluralism is healthy. Moreover, I am convinced that constructive review of most problems can be provided by existing institutions, as when all the different professional and special-interest groups critiqued the Rasmussen study. The acuity of such review hinges on whether the watchers have sufficient competence, resources, and reach.

ELEMENTS OF STEWARDSHIP

Always we will need imaginative, energetic new organizations. But because technical matters are so integral to the modern world, we need at least as much to adapt and reinvigorate the established institutions, such as government and industry, that so inexorably affect our living. And even while we protect research havens and professional-service cloisters, we must not quarantine them from the surrounding social matrix. Panaceas are not even to be hoped for. Changes will have to take place on many fronts, and will have to be tailored to the complexion of the different arenas. In outline, the reforms in attitude, method, and procedure this book has been surveying comprise at least the following, interlocking, elements of stewardship:

- strengthening the basic architectonics of technical trust;
- continually scrutinizing "internal" scientific criteria;
- renewing standards of professional judgment and service;
- sharpening the techniques of analysis for decisionmaking;
- improving the auspices of advising, advocating, and whistleblowing;
- developing societal guidance over procedures of experimentation and over the application of technologies;
- devising structures for structuring medium- to long-term agendas;
- instituting more sensitive and effective means for resolving controversies;
- building-in ways for watching the watchers; and
- constantly striving to orient technical work to societal values.

ENVOI

"The greatest invention of the nineteenth century," Alfred North White-head concluded in *Science and the Modern World*, "was the invention of the method of invention" (1925, p. 136). Truly this was a historic change. Nowadays virtually all innovations are pursued by intention and with strategic method.

The next phase-change was sighted, though a bit wishfully, by J. D. Bernal when, twenty-five years after the publication of his *Social Function of Science*, he pronounced: "The scientific revolution itself has entered a new phase—it has become self-conscious" (1964, p. 209). Technology assessments, values studies, and many other themes of this book will contribute to that self-consciousness.

Although this reflexivity is nowhere near fully realized yet, many lessons have been learned, many monitoring and protective mechanisms have been instituted, and many sensitizations have been achieved. No doubt there will be lapses. But we have every reason to expect fewer exploitive Tuskegee Syphilis Studies, fewer ill-conceived Project Camelots, fewer technocratic excesses like those of the old Atomic Energy Commission, fewer profligate environmental depradations like don't-give-a-damn strip-mining, and fewer gross neglects like that of asbestos. What we will have to be concerned with will be creeping incremental threats like groundwater contamination, environmental degradation brought on by the pressures and temptations of international economic development, a variety of commons problems, occasional high-technology surprises, and tragic strains at the margins of humanitarianism.

The great challenge to our times is to harness research, invention, and professional practice to deliberately embraced human values—to provide *direction* for the directed tragedy of technical progress. Stewardship of a heightened order will be essential. Experts, as I said at the outset, perform both center stage and in the wings. And all of us speak from the citizens' chorus. The fateful questions are how the specialists will interact with the citizens, and whether the performance can be imbued with wisdom, courage, and vision.

Bibliography

Aaron, Henry J., *Politics and the Professors: The Great Society in Perspective* (The Brookings Institution, Washington, D.C., 1978).

Abelson, Philip H., "Chemical abstracts after 75 years," editorial, *Science* 217: 7 (1982).

Abram, Morris B., and Susan M. Wolf, "Public involvement in medical ethics," *New England Journal of Medicine* 310: 627–632 (1984).

"Académie des Sciences: Les chenils de M. Pasteur" [including reply by Pasteur], *Les Annales Politiques et Littéraires* 3: 266 (October 26, 1884).

Achinstein, Peter, *Concepts of Science: A Philosophical Analysis* (The Johns Hopkins University Press, Baltimore, 1968).

Adams, Comfort A., "Cooperation," *Transactions of the American Institute of Electrical Engineers* 38, part 1: 792 (1919).

Allen, Elizabeth [and fifteen others], "Against 'Sociobiology,'" *New York Review of Books*: 43ff. (November 13, 1975); reprinted as 259–268 of Arthur L. Caplan, ed., *The Sociobiology Debate: Readings on Ethical and Scientific Issues* (Harper & Row, New York, 1978).

"Along U.S. coasts: Old solutions fail to solve beach problem," *Geotimes* 26, no. 12: 18–22 (December 1981).

Alonso, William, and Paul Starr, "The political economy of national statistics," *Social Science Research Council Items* 36, no. 3: 29–35 (1982).

Alonso, William, and Paul Starr, eds., *The Political Economy of National Statistics* (Russell Sage Foundation, New York, 1985).

American Association for the Advancement of Science, Committee on Science in the Promotion of Human Welfare, "The Integrity of Science," *American Scientist* 53: 174–198 (1965).

American Association for the Advancement of Science, Committee on Scientific Freedom and Responsibility, *Scientific Freedom and Responsibility* (AAAS, 1515 Massachusetts Ave. NW, Washington, D.C. 20005, 1975).

American Bar Association, Section of Science and Technology, "Curbing ignorance and arrogance: The Science Court proposal and alternatives," *Jurimetrics Journal* 19: 387–459 (1979).

American Chemical Society, Committee on Environmental Improvement and Subcommittee on Environmental Analytic Chemistry, "Guidance for data acquisition and data quality evaluation in environmental chemistry," *Analytical Chemistry* 52: 2242–2249 (1980).

American Industrial Health Council, "AIHC proposal for a science panel" (AIHC, 1075 Central Park Ave., Scarsdale, N.Y. 10583, 1981).

American Institute of Professional Geologists, *Groundwater: Issues and Answers* (AIPG, 7828 Vance Drive, Arvada, Colo. 80003, 1983).

American Joint Committee on Cancer, *Manual for Staging of Cancer* (J. B. Lippincott, Philadelphia, 1983).

American Medical Association, Judicial Council, *Opinions and Reports of the Judicial Council* (AMA, 535 North Dearborn Street, Chicago, Ill. 60610, 1979, 1984).

American Physical Society, "Report to the American Physical Society by the study group on reactor safety," *Reviews of Modern Physics* 47, supplement 1 (1975).

American Psychiatric Association, "American Psychiatric Association statement on the insanity defense" (APA, 1400 K St. NW, Washington, D.C. 20005, 1982).

American Psychiatric Association, Task Force on the Homeless Mentally Ill, *The Homeless Mentally Ill* (APA, 1400 K St. NW, Washington, D.C. 20005, 1984).

American Psychiatric Association and The Hastings Center, special supplement, "In the service of the state: The psychiatrist as double agent," *Hastings Center Report* 7: 1–23 (April 1978).

American Scholar 45, no. 3: 335–359, "Social science: The public disenchantment" (1976).

Anderson, Robert M., Robert Perrucci, Dan E. Schendel, and Leon E. Trachtman, *Divided Loyalties: Whistle-Blowing at BART* (Purdue University, West Lafayette, Ind., 1980).

Andrews, Richard N. L., and Mary Jo Waits, "Theory and methods of environmental values research," *Interdisciplinary Science Review* 5: 71–78 (1980).

Annas, George J., "Life, liberty, and the pursuit of organ sales," *Hastings Center Report* 14, no. 1: 22–23 (1984).

Apostolakis, George, "Bayesian methods in risk assessment," *Advances in Nuclear Science and Technology* 13: 415–465 (1981).

Arrow, Kenneth J., "Rawls's principle of just saving," *Swedish Journal of Economics* 75: 323–335 (1973); reprinted as 133–146 of Kenneth J. Arrow, *Social Choice and Justice* (Harvard University Press, Cambridge, Mass., 1983).

Ascher, William, *Forecasting: An Appraisal for Policy-Makers and Planners* (The Johns Hopkins University Press, Baltimore, 1978).

Ashley, Benedict M., and Kevin D. O'Rourke, *Health Care Ethics: A Theological Analysis*, 2nd ed. (The Catholic Health Association of the United States, 4455 Woodson Rd., St. Louis, Mo. 63134, 1982).

Association of American Medical Colleges, "The maintenance of high ethical standards in the conduct of research" (AAMC, One DuPont Circle NW, Washington, D.C. 20036, June 24, 1982).

Ayyaswamy, P., B. Hauss, T. Hseih, A. Moscati, T. E. Hicks, and D. Okrent, "Estimates of the risks associated with dam failure" (UCLA School of Engineering and Applied Science no. UCLA-ENG-7423, University of California at Los Angeles, Los Angeles, 1974).

Bailar, Barbara A., and C. Michael Lanphier, *Development of Survey Methods to Assess Survey Practices* (American Statistical Association, 806 15th St. NW, Washington, D.C. 20005, 1978).

Baker, Susan P., "Motor vehicle occupant deaths in young children," *Pediatrics* 64: 860–861 (1979).

Baltimore, David, "The Berg letter: Certainly necessary, possibly good," *Hastings Center Report* 10, no. 5: 15 (1980).

Barash, David, *The Whispering Within* (Harper & Row, New York, 1979).

Barber, Bernard, *Science and the Social Order* (Free Press, Glencoe, Ill., 1952).

Barnes, B., and R. G. A. Dolby, "The scientific ethos: A deviant viewpoint," *Archives Européenes de Sociologie* 11: 3–25 (1970).

Barth, Peter S., and H. Allen Hunt, *Workers' Compensation and Work-Related Illnesses and Diseases* (MIT Press, Cambridge, Mass., 1980).

Bartman, Thomas R., "Regulating benzene," 99–134 of Lester B. Lave, ed., *Quantitative Risk Assessment in Regulation* (The Brookings Institution, Washington, D.C., 1982).

Barzun, Jacques, *Science: The Glorious Entertainment* (Harper & Row, New York, 1964).

Batelle Columbus Laboratories, *Interactions of Science and Technology in the Innovative Process: Some Case Studies*, NTIS no. PB228-508 (National Technical Information Service, Springfield, Va., 1973).

Batie, Sandra S., *Soil Erosion: Crisis in America's Croplands?* (The Conservation Foundation, 1717 Massachusetts Ave. NW, Washington, D.C. 20036, 1983).

Bayer, Ronald, *Homosexuality and American Psychiatry: The Politics of Diagnosis* (Basic Books, New York, 1981).

Beauchamp, Tom L., Ruth R. Faden, R. Jay Wallace, Jr., and LeRoy Walters, eds., *Ethical Issues in Social Science Research* (The Johns Hopkins University Press, Baltimore, 1982).

Beauchamp, Tom L., and James F. Childress, *Principles of Biomedical Ethics*, 2nd ed. (Oxford University Press, New York, 1983).

Beckler, David Z., "The precarious life of science in the White House," *Daedalus* 103, no. 3: 115–134 (1974).

Bell, Russell S., and John W. Loop, "The utility and futility of radiographic skull examination for trauma," *New England Journal of Medicine* 284: 236–239 (1971).

Ben David, Joseph, "The ethical responsibility of social scientists: A historical survey and comment," 31–48 of Torgny Segerstedt, ed., *Ethics for Science Policy* (Pergamon Press, New York, 1979).

Bentham, Jeremy, *A Table of the Springs of Action* (Richard and Arthur Taylor, London, 1815); reprinted as 477–512 of Paul McReynolds, ed., *Four Early Works on Motivation* (Scholars' Facsimiles & Reprints, Gainesville, Fla., 1969).

Berg, Paul [and ten others], "Potential biohazards of recombinant DNA molecules," letter to the editor, *Science* 185: 303 (1974).

Berg, Paul, David Baltimore, Sydney Brenner, Richard O. Roblin III, and Maxine F. Singer, "Summary statement of the Asilomar Conference on Recombinant DNA Molecules," *Proceedings of the National Academy of Sciences* 72: 1981–1984 (1975); also printed as "Asilomar Conference on recombinant DNA molecules," *Science* 188: 991–994 (1975).

Berger, T. R., *Northern Frontier—Northern Homeland: Report of the Mackenzie Valley Pipeline Inquiry* (Supply and Services Canada, Ottawa, 1977).

Bernal, J. D., "After twenty-five years," 209–228 of Maurice Goldsmith and Alan MacKay, eds., *Society and Science* (Simon and Schuster, New York, 1964); also printed as preface to the reprinted editon of *The Social Function of Science* (MIT Press, Cambridge, Mass., 1967).

Bernard, Claude, *An Introduction to the Study of Experimental Medicine*, translated by Henry Copley Green (Abelard-Schuman, New York, 1950). Originally published as *Introduction à l'étude de la médecine expérimentale* (J.-B. Ballière, Paris, 1865).

Berwick, Donald M., Shan Cretin, and Emmett B. Keeler, *Cholesterol, Children, and Heart Disease* (Oxford University Press, Oxford, 1980).

Birkhoff, Garrett, *Hydrodynamics: A Study in Logic, Fact and Similitude*, rev. ed. (Princeton University Press, Princeton, N.J., 1960).

Bixler, Gordon, and L. W. Shemilt, eds., *Chemistry and World Food Supplies: The New Frontiers* (International Rice Research Institute, Los Baños, Laguna, Philippines, 1983).

Bloch, Sidney, and Paul Chodoff, *Psychiatric Ethics* (Oxford University Press, New York, 1981).

Bock, Kenneth, *Human Nature and History: A Response to Sociobiology* (Columbia University Press, New York, 1980).

Bode, Hendrik, Frederick Mosteller, John Tukey, and Charles Winsor, "The education of a scientific generalist," *Science* 109: 553–558 (1949).

Boffey, Philip M., "Science court: High officials back test of controversial concept," *Science* 194: 167–169 (1976).

Bonnen, James T., "Federal statistics coordination today: A disaster or a disgrace?" *The American Statistician* 37: 179–192 (1983).

Bordley, James, and A. McGehee Harvey, *Two Centuries of American Medicine 1776–1976* (W. B. Saunders, Philadelphia, 1979).

Brannigan, Vincent, "The first FDA public board of inquiry: The aspartame case," 181–210 of J. D. Nyhart and Milton M. Carrow, eds., *Law and Science in Collaboration* (D. C. Heath, Lexington, Mass., 1983).

Branscomb, Lewis M., "Social science support," letter to the editor, *Science* 213: 1448 (1981).

Braudel, Fernand, *Civilization and Capitalism, 15th–18th Century. Volume II: The Wheels of Commerce*, translated by Siân Reynolds (Harper & Row, New York, 1979).

Breach, Ian, *Windscale Fallout* (Penguin Books, New York, 1978).

Brent, Robert L., "The irresponsible expert witness: A failure of biomedical graduate education and professional accountability," *Pediatrics* 70: 754–762 (1982).

Bridgman, Percy W., "Scientists and social responsibility," *Bulletin of the Atomic Scientists* 4, no. 3: 69–72 (1948).

Bridgman, Percy W., *Reflections of a Physicist* (Philosophical Library, New York, 1950).

Bronowski, Jacob, *Science and Human Values*, rev. ed. (Harper & Row, New York, 1965).

Bronowski, Jacob, "The disestablishment of science," *Encounter* 37: 8–16 (1971).

Brooks, Harvey, "The problem of research priorities," *Daedalus* 107, no. 2: 171–190 (1978).

Brooks, Harvey, "Technology, evolution, and purpose," *Daedalus* 109, no. 1: 65–81 (1980).

Brown, Richard H., *A Poetic for Sociology* (Cambridge University Press, Cambridge, 1977).

Browne, Malcolm W., "In the human equation, risk perceived is risk endured," *New York Times* (March 30, 1980).

Buchanan, Allen E., "Medical paternalism," 61–81 of Rolf Sartorius, ed., *Paternalism* (University of Minnesota Press, Minneapolis, Minn., 1983).

Buckley, Peter, Toksoz Karasu, Edward Charles, and Stefan P. Stein, "Theory and practice in psychotherapy: Some contradictions in expressed belief and practice," *Journal of Nervous and Mental Disease* 167: 218–223 (1979).

Bunker, John P., Benjamin A. Barnes, and Frederick Mosteller, *Costs, Risks, and Benefits of Surgery* (Oxford University Press, New York, 1977).

Burke, John G., "Bursting boilers and the federal power," *Technology and Culture* 7: 1–23 (1966).

Butts, Robert E., "Scientific progress: The Laudan manifesto," *Philosophy of the Social Sciences* 9: 475–483 (1979).

Cahn, Robert, and Patricia L. Cahn, "Lessons from the past," Appendix A to U.S. Council on Environmental Quality and Department of State, *The Global 2000 Report to the President* (U.S. Government Printing Office, Washington, D.C., 1980).

Calabresi, Guido, and Philip Bobbitt, *Tragic Choices* (W. W. Norton, New York, 1978).

Caldwell, Lynton K., *Science and the National Environmental Policy Act* (The University of Alabama Press, University, Ala., 1982).

Callahan, Daniel, "On defining a 'natural death,'" *Hastings Center Report* 7, no. 3: 32–37 (1977).

Callahan, Daniel, and Bruce Jennings, eds., *Ethics, the Social Sciences, and Policy Analysis* (Plenum Press, New York, 1983).

Campbell, Donald T., "On the conflicts between biological and social evolution and between psychology and moral tradition," *American Psychologist* 30: 1103–1126 (1975).

Campbell, Gregory L., David Cohan, and D. Warner North, "The application of decision analysis to toxic substances: Proposed methodology and two case studies," NTIS no. PB82-249-103 (National Technical Information Service, Springfield, Va., 1982).

Caplan, Arthur L., ed., *The Sociobiology Debate: Readings on Ethical and Scientific Issues* (Harper & Row, New York, 1978).

Carson, Rachel, *Silent Spring* (Houghton Mifflin, Boston, 1962).

Cartwright, Nancy, *How the Laws of Physics Lie* (Oxford University Press, New York, 1983).

Casper, Barry M., "Technology policy and democracy," *Science* 194: 29–35 (1976).

Casper, Barry M., "Laser enrichment: A new path to proliferation?" *Bulletin of the Atomic Scientists* 33: 28–41 (1977).

Cassel, Christine, and Andrew Jameton, "Medical responsibility and thermonuclear war," 9–23 of Christine Cassel, Michael McCally, and Henry Abraham, eds., *Nuclear Weapons and Nuclear War* (Praeger Publishers, New York, 1984).

Cellarius, Richard A., and John Platt, "Councils of urgent studies," *Science* 177: 670–676 (1972).

Chagnon, Napoleon A., and William Irons, eds., *Evolutionary Biology and Human Social Behavior: An Anthropological Perspective* (Duxbury Press, North Scituate, Mass., 1979).

Chalk, Rosemary, and Frank von Hippel, "Due process for dissenting 'whistle-blowers,'" *Technology Review* 81: 49–55 (1979).

Chalk, Rosemary, Mark S. Frankel, and Sallie B. Chafer, *AAAS Professional Ethics Project: Professional Ethics Activities in the Scientific and Engineering Societies* (American Association for the Advancement of Science, 1515 Massachusetts Ave. NW, Washington, D.C. 20005, 1980).

Chandler, William U., "Improving world health: A least cost strategy," Worldwatch Paper no. 59 (Worldwatch Institute, 1776 Massachusetts Ave. NW, Washington, D.C. 20036, 1984).

Christians, Clifford C., and Jay M. Van Hook, eds., *Jacques Ellul: Interpretive Essays* (University of Illinois Press, Urbana, Ill., 1981).

Clark, Stephen R. L., *The Moral Status of Animals* (Oxford University Press, Oxford, 1977).

Coates, Vary T., and Bernard Finn, *A Retrospective Technology Assessment: Submarine Telegraphy* (San Francisco Press, San Francisco, 1979).

Cochran, Thomas B., and David L. Bodde, "Conflicting views on a neutrality criterion for radioactive waste management," 110–128 of Douglas MacLean and Peter G. Brown, eds., *Energy and the Future* (Rowman and Littlefield, Totowa, N.J., 1983).

Cochran, William G., Frederick Mosteller, and John W. Tukey, *Statistical Problems of the Kinsey Report* (American Statistical Association, Washington, D.C., 1954).

Cohen, Morris L., Naomi Ronen, and Jan Stepan, *Law & Science: A selected bibliography* (MIT Press, Cambridge, Mass., 1980).

Cole, Gerald A., and Stephen B. Withey, "Perspectives on risk perceptions," *Risk Analysis* 1: 143–163 (1981).

Coleman, James S., *Equality of Educational Opportunity* (U.S. Government Printing Office, Washington, D.C., 1966).

Colton, Kent W., and Kenneth L. Kraemer, "Technology, society, and public policy: EFT [electronic funds transfer] systems," 241–261 of David M. O'Brien and Donald A. Marchand, eds., *The Politics of Technology Assessment* (Lexington Books, Lexington, Mass., 1982).

Comroe, Julius H., Jr., and Robert D. Dripps, "Scientific basis for the support of biomedical science," *Science* 192: 105–111 (1976).

Cooney, William, "Medical group sues state over mental health program," *San Francisco Chronicle* (November 16, 1979).

Corner, George W., *A History of The Rockefeller Institute, 1901–1953* (The Rockefeller Institute Press, New York, 1964).

Council for Science and Society, *The Acceptability of Risks* (Council for Science and Society, 3/4 St. Andrew's Hill, London EC4 5BY, 1977).

Council for Science and Society and the Governors of the British Institute of Human Rights, *Scholarly Freedom and Human Rights* (Council for Science and Society, 3/4 St. Andrew's Hill, London EC4 5BY, 1977).

Cousins, Norman, "The fallacy of cost–benefit ratio," *Saturday Review* 6, no. 8: 8 (1979).

Covello, Vincent T., W. Gary Flamm, Joseph V. Rodricks, and Robert G. Tardiff, eds., *The Analysis of Actual Versus Perceived Risks* (Plenum Press, New York, 1983).

Cowan, Ruth Schwartz, *More Work for Mother: The Ironies of Household Technology from the Open Hearth to the Microwave Oven* (Basic Books, New York, 1983).

Crane, Diana, *Invisible Colleges: Diffusion of Knowledge in Scientific Communities* (The University of Chicago Press, Chicago, 1972).

Cronbach, Lee J., "Prudent aspirations for social inquiry," 61–81 of William H. Kruskal, ed., *The Social Sciences: Their Nature and Uses* (The University of Chicago Press, Chicago, 1982).

Cullen, Michael J., *The Statistical Movement in Early Victorian Britain: The Foundations of Empirical Social Research* (Harvester Press, Hassocks, Sussex, 1975).

Dahl, Robert A., *A Preface to Democratic Theory* (The University of Chicago Press, Chicago, 1956).

Dallmayr, Fred R., and Thomas A. McCarthy, eds., *Understanding and Social Inquiry* (University of Notre Dame Press, Notre Dame, Ind., 1977).

Daniels, Norman, *Reading Rawls* (Basic Books, New York, 1976).

Darwin, Charles, British Museum manuscript collection, additional manuscript 37725, f.6 (1861).

DeBeer, Gavin, *The Sciences Were Never At War* (Thomas Nelson and Sons, London, 1960).

Delgado, Richard, and David R. Millen, "God, Galileo, and government: Toward constitutional protection for scientific inquiry," 53 *Washington Law Review*: 349–404 (1978).

Delgado, Richard, and Peter McAllen, "The moralist as expert witness," 62 *Boston University Law Review*: 869–926 (1982).

Diamond, Stuart, "Credit file password is stolen," *New York Times* (June 23, 1984).

Dickson, Edward M., and Raymond Bowers, *The Video Telephone: A New Era in Telecommunications* (Praeger Publishers, New York, 1974).

DiSanti, Richard J., "Cost–benefit analysis for standards regulating toxic substances under the Occupational Safety and Health Act: *American Petroleum Institute* v. *OSHA*," 60 *Boston University Law Review*: 115–141 (1980).

Dixon, Bernard, "Engineering chimeras for Noah's ark," *Hastings Center Report* 14, no. 2: 10–12 (1984).

Doll, Richard, and Richard Peto, *The Causes of Cancer* (Oxford University Press, Oxford, 1981).

Doppelt, Gerald, "Kuhn's epistemological relativism: An interpretation and defense," *Inquiry* 21: 33–86 (1978).

Doppelt, Gerald, "Relativism and recent pragmatic conceptions of scientific rationality," 107–142 of Nicholas Rescher, ed., *Scientific Explanation and Understanding* (University Press of America, Lanham, Md., 1983).

Dubos, René, "The despairing optimist," *American Scholar* 42, no. 4: 547–549 (1973).

Duffy, John, *The Healers: A History of American Medicine* (McGraw-Hill, New York, 1976).

Dworkin, Gerald, "Paternalism," 107–126 of Richard A. Wasserstrom, ed., *Morality and the Law* (Wadsworth Publishers, Belmont, Calif., 1971).

Dworkin, Gerald, "Paternalism: Some second thoughts," 105–111 of Rolf Sartorius, ed., *Paternalism* (University of Minnesota Press, Minneapolis, Minn., 1983).

Dworkin, Ronald, *Taking Rights Seriously*, 6th printing (Harvard University Press, Cambridge, Mass., 1977).

Dyson, Freeman J., "Death of a project," *Science* 149: 141–144 (1965).

Eddy, David, *Screening for Cancer: Theory, Analysis, and Design* (Prentice-Hall, Englewood Cliffs, N.J., 1980).

Edel, Abraham, *Ethical Judgment: The Use of Science in Ethics* (Free Press, New York, 1955).

Edel, Abraham, "Some current trends in ethical theory," 9–28 of Daniel Jeremy Silver, ed., *Judaism and Ethics* (Ktav Publishing House, New York, 1970).

Edsall, John T., *Scientific Freedom and Responsibility* (American Association for the Advancement of Science, Washington, D.C., 1975); an abbreviated version appeared as "Scientific freedom and responsibility," *Science* 188: 687–693 (1975).

Ehrlich, Paul R., *The Population Bomb* (Ballantine, New York, 1968).

Eisenstadt, S. N., with M. Curelaru, *The Form of Sociology—Paradigms and Crises* (John Wiley & Sons, New York, 1976).

Eisner, Thomas, "AAAS human rights activities," letter to the editor, *Science* 222: 6–8 (1983).

Electric Power Research Institute, "Critique of the AEC reactor safety study (Wash-1400)," EPRI no. 217-2-3 (EPRI, 3412 Hillview Ave., Palo Alto, Calif. 94303, 1975).

Ellul, Jacques, *The Technological Society*, translated by John Wilkinson (Vintage, New York, 1964). Originally published as *La Technique ou l'enjeu du siècle* (Armand Colin, Paris, 1954).

Elster, Jon, *Ulysses and the Sirens* (Cambridge University Press, Cambridge, 1979).

Environmental Health Perspectives 58: 322–381, "Consensus report on formaldehyde" (1984).

Evelyn, John, *Sylva, or A Discourse of forest-trees, and the propagation of timber in His Majesties dominions* (John Martyn & James Allestry, London, 1664); available in microprint facsimile (Readex Microprint, New York, 1970).

Fein, Rashi, "On measuring economic benefits of health programs," 181–217 of Gordon McLachlan and Thomas McKeown, eds., *Medical History and Medical Care* (Oxford University Press, Oxford, 1971).

Feinstein, Alvan R., *Clinical Judgment* (Williams & Wilkins, Baltimore, 1967).

Ferrando, R., *Traditional and Non-Traditional Foods* ([United Nations] Food and Agriculture Organization, Rome, 1981).

Fischhoff, Baruch, "Cost benefit analysis and the art of motorcycle maintenance," *Policy Sciences* 8: 177–202 (1977).

Fischhoff, Baruch, Sarah Lichtenstein, Paul Slovic, Stephen L. Derby, and Ralph L. Keeney, *Acceptable Risk* (Cambridge University Press, New York, 1981).

Fishkin, James S., ed., "Symposium on the theory and practice of representation," *Ethics* 91: 353–490 (1981).

Fletcher, John C., "Emerging ethical issues in fetal therapy," 293–318 of Kåre Berg and Knut Eric Tranøy, eds., *Research Ethics* [vol. 128 of *Progress in Clinical and Biological Research*] (Alan R. Liss, New York, 1983).

Flew, Antony, *Evolutionary Ethics* (Macmillan, London, 1967).

Flexner, Abraham, *Medical Education in the United States and Canada; A Report to the Carnegie Foundation for the Advancement of Teaching* (Carnegie Foundation, New York, 1910; reprinted by Arno Press, New York, 1972).

Florman, Samuel C., *The Existential Pleasures of Engineering* (St. Martin's Press, New York, 1976).

Florman, Samuel C., *Blaming Technology: The Irrational Search for Scapegoats* (St. Martins's Press, New York, 1981).

Food Safety Council, *Proposed System for Food Safety Assessment: Final Report of the Scientific Committee of the Food Safety Council* (FSC, 1725 K St. NW, Washington, D.C. 20006, 1980).

Food Safety Council, *A Proposed Food Safety Evaluation Process* (FSC, 1725 K St. NW, Washington, D.C. 20006, 1982).

Ford, Daniel, *The Cult of the Atom: The Secret Papers of the Atomic Energy Commission* (Simon and Schuster, New York, 1982).

Foss, Dennis C., *The Value Controversy in Sociology* (Jossey-Bass, San Francisco, 1977).

Fournet, Jacques, *Plants and Flowers of the Caribbean*, translated by Deva Tirvengadum (Éditions du Pacifique, Papeete, Tahiti, 1977).

Fox, Renée C., "The medicalization and demedicalization of American society," *Daedalus* 106, no. 1: 9–22 (1977); reprinted as 465–483 of Renée C. Fox, *Essays in Medical Sociology* (John Wiley & Sons, New York, 1979).

Franck, Isaac, guest ed., special issue on "Medical ethics from the Jewish perspective," *Journal of Medicine and Philosophy* 8, no. 3: 207–328 (1983).

Freedman, Daniel G., *Human Sociobiology: A Holistic Approach* (Free Press, New York, 1979).

Freidson, Eliot, *The Profession of Medicine* (Dodd, Mead, New York, 1970).

French, Peter A., "Collective responsibility and the practice of medicine," *Journal of Medicine and Philosophy* 7: 65–85 (1982).

Friedman, Milton, *Essays in Positive Economics* (The University of Chicago Press, Chicago, 1953).

Galston, Arthur W., "Science and social responsibility: A case history," *Annals of the New York Academy of Sciences* 196: 223–235 (1972).

Galston, William A., *Justice and the Human Good* (The University of Chicago Press, Chicago, 1980).

Gamble, D. J., "The Berger inquiry: An impact assessment process," *Science* 199: 946–952 (1978).

Gaylin, Willard, editor, "The XYY controversy: Research into violence and genetics," special supplement to *Hastings Center Report* 9 (August 1980).

Geballe, Theodore H., "This golden age of solid-state physics," *Physics Today* 34: 132–142 (1981).

Gendron, Bernard, *Technology and the Human Condition* (St. Martin's Press, New York, 1977).

Gergen, Kenneth J., *Toward Transformation in Social Knowledge* (Springer-Verlag, New York, 1982).

Geuss, Raymond, *The Idea of a Critical Theory: Habermas and the Frankfurt School* (Cambridge University Press, Cambridge, 1981).

Gibson, H. B., *Pain and its Conquest* (Peter Owen, London, 1982).

Giedion, Siegfried, *Mechanization Takes Command* (Oxford University Press, New York, 1948).

Gilpin, Robert, and Christopher Wright, eds., *Scientists and National Policy-Making* (Columbia University Press, New York, 1964).

Glazer, Myron, "Ten whistleblowers and how they fared," *Hastings Center Report* 13, no. 6: 33–41 (1983).

Glen, William, *The Road to Jaramillo: Critical Years of the Revolution in Earth Science* (Stanford University Press, Stanford, Calif., 1982).

Golden, William T., guest ed., special issue on "Science advice to the President," *Technology in Society* 2, no. 1–2: 1–256 (1980).

Gorbach, Sherwood L., ed., "Risk assessment of recombinant DNA experimentation with *Escherichia coli* K12," *Journal of Infectious Diseases* 137: 612–714 (1978).

Gordon, Wendy, *A Citizen's Handbook on Groundwater Protection* (Natural Resources Defense Council, 122 East 42nd St., New York, N.Y. 10168, 1984).

Goudie, Andrew, *The Human Impact: Man's Role in Environmental Change* (MIT Press, Cambridge, Mass., 1981).

Gould, Stephen Jay, *Ever Since Darwin: Reflections in Natural History* (W. W. Norton, New York, 1977).

Gould, Stephen Jay, *The Mismeasure of Man* (W. W. Norton, New York, 1981a).

Gould, Stephen Jay, "Evolution as fact and theory," *Discover* 2, no. 5: 35ff. (1981b).

Graham, John D., and James W. Vaupel, "Value of a life: What difference does it make?" *Risk Analysis* 1: 89–95 (1981).

Graham, Loren R., "Concerns about science and attempts to regulate inquiry," *Daedalus* 107, no. 2: 1–21 (1978).

Graham, Loren R., *Between Science and Values* (Columbia University Press, New York, 1981).

Green, Melvin R., "Evolution of ASME voluntary standards and their future," *Mechanical Engineering* 101: 106–108 (1979).

Greenberger, Martin, *Caught Unawares: The Energy Decade in Retrospect* (Ballinger, Cambridge, Mass., 1983).

Gregory, Michael S., Anita Silvers, and Diane Sutch, eds., *Sociobiology and Human Nature: An Interdisciplinary Critique and Defense* (Jossey-Bass, San Francisco, 1978).

Griesmeyer, J. M., M. Simpson, and D. Okrent, *The Use of Risk Aversion in Risk Acceptance Criteria* (UCLA School of Engineering and Applied Science no. UCLA-ENG-7970, University of California at Los Angeles, Los Angeles, 1979).

Gustafson, James M., "Theology confronts technology and the life sciences," *Commonweal* 105: 386–392 (1978).

Gutmann, Amy, *Liberal Equality* (Cambridge University Press, Cambridge, 1980).

Gutting, Gary, ed., *Paradigms and Revolutions: Appraisals and Applications of Thomas Kuhn's Philosophy of Science* (University of Notre Dame Press, Notre Dame, Ind., 1980).

Haan, Norma, Robert N. Bellah, Paul Rabinow, and William M. Sullivan, eds., *Social Science As Moral Inquiry* (Columbia University Press, New York, 1983).

Hagstrom, Warren O., *The Scientific Community* (Basic Books, New York, 1965).

Haimes, Yacov Y., *Risk/Benefit Analysis in Water Resources Planning and Management* (Plenum Press, New York, 1981).

Halévy, Élie, *La Jeunesse de Bentham* (Ancienne Librairie Germer Baillière, Paris, 1901).

Hallam, Anthony, *Great Geological Controversies* (Oxford University Press, Oxford, 1983).

Hampshire, Stuart, "The illusion of sociobiology" [a review of Edward O. Wilson's *On Human Nature*], *New York Review of Books* 25, no. 15: 64–69 (1978).

Handler, Philip, "Search for truth?" interview, *News Report of the National Academy of Sciences* 19: 9 (Washington, D.C., March 1969).

Handler, Philip, "Of questions and committees," *The National Research Council in 1977*: 3–20 (National Academy of Sciences, Washington, D.C., 1978).

Handler, Philip, "On reports," *The National Research Council in 1978*: 3–29 (National Academy of Sciences, Washington, D.C., 1979).

Hanft, Ruth A., "Use of social science data for policy analysis and policymaking," 249–270 of Daniel Callahan and Bruce Jennings, eds., *Ethics, the Social Sciences, and Policy Analysis* (Plenum Press, New York, 1983).

Hanson, Norwood Russell, *Patterns of Discovery* (Cambridge University Press, Cambridge, 1958).

Hart, H. L. A., *Punishment and Responsibility* (Oxford University Press, Oxford, 1968).

Harvard Educational Review, *Equal Educational Opportunity* (Harvard University Press, Cambridge, Mass., 1969).

Haydon, Graham, "On being responsible," *Philosophical Quarterly* 28: 46–57 (1978).

Heilbroner, Robert, "Do machines make history?" *Technology and Culture* 8: 333–345 (1967).

Heilbroner, Robert L., "Economics as a 'value-free' science," *Social Research* 40: 129–143 (1973).

Heims, Steve J., *John von Neumann and Norbert Wiener* (MIT Press, Cambridge, Mass., 1980).

Herschel, John Frederick William, untitled essay on Adolphe Quételet's recently translated *Theory of Probabilities*, *The Edinburgh Review* 92: 1–57 (1850).

High, Dallas M., "Is 'natural death' an illusion?" *Hastings Center Report* 8, no. 4: 37–42 (1978).

Hill, Austin Bradford, "The environment and disease: Association or causation?" *Proceedings of the Royal Society of Medicine* 58: 295–300 (1965).

Hinshaw, Robert E., "Anthropology, administration, and public policy," *Annual Review of Anthropology* 9: 497–522 (1980).

Hofstadter, Richard, *Social Darwinism and American Thought* (University of Pennsylvania Press, Philadelphia, 1944; rev. ed., Braziller, New York, 1959).

Hohenemser, Christoph, "Summary of panel discussion and commentary," 49–67 of Vincent T. Covello, W. Gary Flamm, Joseph V. Rodricks, and Robert G. Tardiff, eds., *The Analysis of Actual Versus Perceived Risks* (Plenum Press, New York, 1983).

Holden, Constance, "Reagan versus the social sciences," *Science* 226: 1052–1054 (1984).

Holling, C. S., ed., *Adaptive Environmental Assessment and Management* (John Wiley & Sons, New York, 1978).

Holton, Gerald, and Robert S. Morison, editors, *Limits of Scientific Inquiry* (W. W. Norton, New York, 1979; this is a reprint of *Daedalus* 107, no. 2, 1978).

Horowitz, Irving Louis, *The Rise and Fall of Project Camelot* (MIT Press, Cambridge, Mass., 1967).

Horwitch, Mel, *Clipped Wings: The American SST Conflict* (MIT Press, Cambridge, Mass., 1982).

Howe, Adolph, *On Economic Knowledge: Toward a Science of Political Economics* (M. S. Sharpe, Armonk, N.Y., 1977).

Huff, Toby E., "Discovery and explanation in sociology: Durkheim on suicide," *Philosophy of the Social Sciences* 5: 241–257 (1975).

Hughes, J. Donald, *Ecology in Ancient Civilizations* (University of New Mexico Press, Albuquerque, N.M., 1975).

Hughes, Thomas P., *Networks of Power: Electrification in Western Society, 1880–1930* (The Johns Hopkins University Press, Baltimore, 1983).

Hull, David L., "Laudan's Progress and its Problems," *Philosophy of the Social Sciences* 9: 457–465 (1979).

Humphreys, Laud, *Tearoom Trade: Impersonal Sex in Public Places*, rev. ed. (Aldine Publishing, Hawthorne, N.Y., 1975).

Hunt, James H., *Selected Readings in Sociobiology* (McGraw-Hill, New York, 1980).

Hutt, Peter Barton, "The use of an advisory commission," 51 *Southern California Law Review*: 1435–1469 (1978).

Huxley, T. H., and Julian Huxley, *Evolution and Ethics 1893–1943* (Pilot Press, London, 1947).

Hyams, Edward, *Soil and Civilization* (Harper & Row, New York, 1976).

Illinois Institute of Technology Research Institute, *Technology in Retrospect and Critical Events in Science* [*TRACES*], NTIS no. PB-234 767/2 (National Technical Information Service, Springfield, Va., December 15, 1968).

Inglehart, Ronald, *The Silent Revolution: Changing Values and Political Styles Among Western Publics* (Princeton University Press, Princeton, N.J., 1977).

Inhaber, Herbert, *Energy Risk Assessment* (Gordon and Breach, New York, 1982).

Institute of Medicine, *Behavioral Science and the Secret Service: Toward the Prevention of Assassination* (National Academy Press, Washington, D.C., 1981).

Institute of Medicine, Committee for a Planning Study for an Ongoing Study of Costs of Environment-Related Health Effects, *Costs of Environment-Related Health Effects: A Plan for Continuing Study* (National Academy Press, Washington, D.C., 1981).

Institute of Medicine, Committee for a Study of the Health Care of Racial/Ethnic Minorities and Handicapped Persons, *Health Care in a Context of Civil Rights* (National Academy Press, Washington, D.C., 1981).

Institute of Medicine, Committee on Issues and Priorities for New Vaccine Development, *New Vaccine Development: Establishing Priorities. Volume I: Diseases of Importance to the United States* (National Academy Press, Washington, D.C., 1984).

Institute of Medicine/National Research Council, Committee for a Study on Saccharin and Food Safety Policy, *Food Safety Policy: Scientific and Societal Considerations* (National Academy of Sciences, Washington, D.C., 1979).

Insurance Institute for Highway Safety, *Children in Crashes* (IIHS, Watergate 600, Washington, D.C. 20037, 1981).

Jackson, David A., and Stephen P. Stich, eds., *The Recombinant DNA Debate* (Prentice-Hall, Englewood Cliffs, N.J., 1979).

Jarvie, I. C., "Laudan's problematic progress and the social sciences," *Philosophy of the Social Sciences* 9: 484–497 (1979).

Jensen, Allan Astrup, "Chemical contaminants in human milk," *Residue Reviews* 89: 1–128 (1983).

Jevons, W. Stanley, *The Coal Question; An Inquiry Concerning the Progress of the Nation, and the Probable Exhaustion of Our Coal-Mines* (Macmillan, London, 1865).

Jonas, Hans, *The Imperative of Responsibility: In Search of an Ethics for the Technological Age* (The University of Chicago Press, Chicago, 1984).

Jones, Greta, *Social Darwinism and English Thought: The Interaction Between Biological and Social Theory* (Humanities Press, Atlantic Highlands, N.J., 1980).

Jones-Lee, M. W., *The Value of Life* (The University of Chicago Press, Chicago, 1976).

Jungk, Robert, *Brighter than a Thousand Suns: A Personal History of the Atomic Scientists*, translated by James Cleugh (Harcourt, Brace & World, New York, 1958).

Kahneman, Daniel, Paul Slovic, and Amos Tversky, *Judgment Under Uncertainty: Heuristics and Biases* (Cambridge University Press, Cambridge, 1982).

Kakar, Sudhir, *Frederick Taylor: A Study in Personality and Innovation* (MIT Press, Cambridge, Mass., 1970).

Kantrowitz, Arthur, "Proposal for an institution for scientific judgment," *Science* 156: 763–764 (1967).

Kantrowitz, Arthur, "Controlling technology democratically," *American Scientist* 63: 505–509 (1975).

Kantrowitz, Arthur, "The science court experiment," *Jurimetrics* 17: 332–341 (1977).

Kaplan, Abraham, "Philosophy of science in anthropology," *Annual Review of Anthropology* 13: 25–39 (1984).

Karl, Barry D., "The social sciences and Mr. Hoover: Recent Social Trends," 201–225 of his *Charles E. Merriam and the Study of Politics* (The University of Chicago Press, Chicago, 1974).

Karrh, Bruce W., Thomas W. Carmody, Robert M. Clyne, Kenneth G. Gould, Gloria Portela-Cubria, Jerry M. Smith, and Milton Freifeld, "Guidance for the evaluation, risk assessment and control of chemical embryo-fetotoxins," *Journal of Occupational Medicine* 23: 397–399 (1981).

Kaufman, Martin, *American Medical Education: The Formative Years, 1765–1910* (Greenwood Press, Westport, Conn., 1976).

Keeney, Ralph L., and Howard Raiffa, *Decisions with Multiple Objectives: Preferences and Value Tradeoffs* (John Wiley & Sons, New York, 1976).

Keller, Suzanne, *Beyond the Ruling Class: Strategic Elites in Modern Society* (Random House, New York, 1963).

Kelly, Alfred, *The Descent of Darwin: The Popularization of Darwinism in Germany, 1860–1914* (University of North Carolina Press, Chapel Hill, N.C., 1981).

Kennedy, John F., press conference, May 22, 1962, *The Public Papers of the Presidents of the United States: John F. Kennedy, 1962*: 422 (U.S. Government Printing Office, Washington, D.C., 1963).

Keohane, Robert O., and J. S. Nye, "Organizing for global environmental and resource interdependence," 46–63 of Appendix B to Volume I of U.S. Commission on Organization of the Government for the Conduct of Foreign Policy, *Report* (U.S. Government Printing Office, Washington, D.C., 1975).

Kerr, Richard A., "Whither the shoreline?" *Science* 214: 428 (1982).

Key, Marcus M., Lorin E. Kerr, and Merle Bundy, eds., *Pulmonary Reactions to Coal Dust* (Academic Press, New York, 1971).

Keyworth, George A., II, "APS steps into a political vortex," *Physics Today* 36, no. 5: 8 (1983).

Kitts, David B., *The Structure of Geology* (Southern Methodist University Press, Dallas, 1977).

Klaw, Spencer, *The New Brahmins: Scientific Life in America* (William Morrow, New York, 1968).

Kleinig, John, *Paternalism* (Rowman & Allanheld, Totowa, N.J., 1983).

Kluckhohn, Clyde, "Have there been discernable shifts in American values during the past generation?" 145–217 of Elting E. Morison, ed., *The American Style* (Harper & Brothers, New York, 1958).

Kluckhohn, Clyde, et al., "Values and value-orientations in the theory of action," 388–433 of Talcott Parsons and Edward A. Shils, eds., *Toward a General Theory of Action* (Harvard University Press, Cambridge, Mass., 1967).

Kluckhohn, Florence Rockwood, and Fred L. Strodtbeck, *Variations in Value Orientations* (Row, Peterson, Evanston, Ill., 1961; reprinted by Greenwood Press, Westport, Conn., 1973).

Kohlstedt, Sally G., *The Formation of the American Scientific Community* (University of Illinois Press, Urbana, Ill., 1976).

Kopp, Carolyn, "The origins of the American scientific debate over fallout hazards," *Social Studies of Science* 9: 403–422 (1979).

Krimsky, Sheldon, *Genetic Alchemy: The Social History of the Recombinant DNA Controversy* (MIT Press, Cambridge, Mass., 1982).

Kuhn, Thomas, *The Structure of Scientific Revolutions*, 2nd ed. (The University of Chicago Press, Chicago, 1970).

Lakatos, Imré, and Alan Musgrave, eds., *Criticism and the Growth of Knowledge* (Cambridge University Press, Cambridge, 1970).

Lambright, W. Henry, *Shooting Down the Nuclear Plane* (Inter-University Case Program and Bobbs-Merrill, Indianapolis, Ind., 1967).

Landefeld, J. Steven, and Eugene P. Seskin, "The economic value of life: Linking theory to practice," *American Journal of Public Health* 72: 555–566 (1982).

Landes, David S., *The Unbound Prometheus: Technological Change and Industrial Development in Western Europe from 1750 to the Present* (Cambridge University Press, Cambridge, 1969).

Lang, Serge, *The File: A Case Study in Correction (1977–1979)* (Springer-Verlag, New York, 1981).

Lapp, Ralph E., *The New Priesthood: The Scientific Elite and the Uses of Power* (Harper & Row, New York, 1965).

Lappé, Marc, "Ethical issues in testing for differential sensitivity to occupational hazards," *Journal of Occupational Medicine* 25: 797–808 (1983).

Latour, Bruno, and Steve Woolgar, *Laboratory Life: The Social Construction of Scientific Facts* (Sage Publications, Beverly Hills, Calif., 1979).

Laudan, Larry, *Progress and its Problems: Toward a Theory of Scientific Growth* (University of California Press, Berkeley, Calif., 1977).

Laudan, Larry, *Mind and Medicine: Problems of Explanation and Evaluation in Psychiatry and the Biomedical Sciences* (University of California Press, Berkeley, Calif., 1982).

Lave, Lester B., "Risk assessment for regulation of dioxin," 635–638 of Richard E. Tucker, Alvin L. Young, and Allan P. Gray, eds., *Human and Environmental Risks of Chlorinated Dioxins and Related Compounds* (Plenum Press, New York, 1983).

Layton, Edwin T., Jr., *The Revolt of the Engineers: Social Responsibility and the American Engineering Profession* (Case Western Reserve University Press, Cleveland, Ohio, 1971).

Lebacqz, Karen, ed., *Genetics, Ethics and Parenthood* (Pilgrim Press, New York, 1983).

Leeds, Anthony, and Valentine Dusek, eds., special double issue on "Sociobiology: The debate evolves," *The Philosophical Forum* 13, no. 2–3: 1–323 (1981–1982).

Leopold, Aldo, *Sand County Almanac* (Oxford University Press, New York, 1949).

Levi, Isaac, *The Enterprise of Knowledge: An Essay on Knowledge, Credal Probability, and Chance* (MIT Press, Cambridge, Mass., 1980).

Lewis, Harold W., "The safety of fission reactors," *Scientific American* 242: 53–65 (1980).

Lindblom, Charles E., *Politics and Markets* (Basic Books, New York, 1977).

Lippmann, Walter, *Drift and Mastery* (M. Kennerley, New York, 1914).

Lippmann, Walter, *Essays in the Public Philosophy* (Little, Brown, Boston, 1955).

Lively, Jack, "Paternalism," 147–165 of A. Phillips Griffiths, ed., *Of Liberty* (Cambridge University Press, Cambridge, 1983).

Lloyd, Christopher, ed., *Social Theory and Political Practice* (Oxford University Press, Oxford, 1983).

Lovins, Amory B., *Soft Energy Paths: Toward a Durable Peace* (Ballinger, Cambridge, Mass., 1977).

Lowrance, William W., *Of Acceptable Risk: Science and the Determination of Safety* (William Kaufmann, Los Altos, Calif., 1976).

Lowrance, William W., "The NAS surveys of fundamental research 1962–1974, in retrospect," *Science* 197: 1254–1260 (1977).

Lowrance, William W., ed., *Assessment of Health Effects at Chemical Disposal Sites* (William Kaufmann, Los Altos, Calif., 1981).

Lowrance, William W., "Choosing our pleasures and our poisons: Risk assessment for the 1980s," 99–130 of Albert H. Teich and Ray Thornton, eds., *Science, Technology, and the Issues of the Eighties: Policy Outlook* (Westview Press, Boulder, Colo., 1982a).

Lowrance, William W., "Watching the watchers," letter to the editor, *Science* 216: 1172 (1982b).

Lowrance, William W., ed., *Public Health Risks of the Dioxins* (William Kaufmann, Los Altos, Calif., 1984).

Lugg, Andrew, "Laudan and the problem-solving approach to scientific progress and rationality," *Philosophy of the Social Sciences* 9: 466–474 (1979).

Lumsden, Charles J., and Edward O. Wilson, *Genes, Mind, and Culture: The Coevolutionary Process* (Harvard University Press, Cambridge, Mass., 1981).

Luria, Salvador E., "Directed genetic change: Perspectives from molecular genetics," 1–19 of T. M. Sonnenborn, ed., *The Control of Human Heredity and Evolution* (Macmillan, New York, 1965).

Lyons, Henry, *The Royal Society 1660–1940* (Cambridge University Press, Cambridge, 1944).

MacLean, Douglas, and Peter G. Brown, eds., *Energy and the Future* (Rowman and Littlefield, Totowa, N.J., 1983).

MacLeod, Roy, and Peter Collins, eds., *The Parliament of Science: The British Association for the Advancement of Science 1831–1981* (Science Reviews Ltd., 40, The Fairway, Northwood, Middlesex HA6 3DY, 1981).

Malthus, Thomas Robert, *An Essay on the Principle of Population, a New Edition* (printed for J. Johnson by T. Bensley, London, 1803).

Margenau, Henry, *Ethics and Science* (Van Nostrand, Princeton, N.J., 1964).

Markle, Gerald E., and James C. Peterson, eds., *The Laetrile Phenomenon* (Westview Press, Boulder, Colo., 1980).

Marshak, Robert E., "APS and public issues," *Physics Today* 36, no. 5: 9 (1983).

Martin, James, "The proposed 'science court,'" 75 *Michigan Law Review*: 1058–1091 (1977).

Martin, Rex, and James W. Nickel, "Recent work on the concept of rights," *American Philosophical Quarterly* 17: 165–180 (1980).

Marx, Karl, *Capital*, edited by Frederick [sic] Engels (English reprint of the German edition of 1887, Foreign Languages Publishing House, Moscow, 1961).

Mattern, Ruth, "Altruism, ethics, and sociobiology," 462–475 of Arthur L. Caplan, ed., *The Sociobiology Debate: Readings on Ethical and Scientific Issues* (Harper & Row, New York, 1978).

Mazur, Allan, "Disputes between experts," *Minerva* 11: 243–262 (1973).

Mazur, Allan, "Science courts," *Minerva* 15: 1–14 (1977).

McAllister, Donald M., *Evaluation in Environmental Planning: Assessing Environmental, Social, Economic, and Political Tradeoffs* (MIT Press, Cambridge, Mass., 1980).

McCormick, Norman J., *Reliability and Risk Analysis* (Academic Press, New York, 1981).

McCormick, Richard A., *How Brave a New World? Dilemmas in Bioethics* (Doubleday, New York, 1981).

McCormick, Richard A., *Health and Medicine in the Catholic Tradition* (Crossroad, New York, 1984).

McGarity, Thomas O., "Substantive and procedural discretion in administrative resolution of science policy questions: Regulating carcinogens in EPA and OSHA," 67 *Georgetown Law Journal*: 729–810 (1979).

McGinn, Robert E., "What is technology?" 179–197 of Paul T. Durbin, ed., *Research in Philosophy and Technology* 1 (Jai Press, Greenwich, Conn., 1978).

McKenzie, D. P., "Plate tectonics and its relationship to the evolution of ideas in the geological sciences," *Daedalus* 106, no. 3: 97–124 (1977).

McNeil, Barbara J., Ralph Weichselbaum, and Stephen G. Pauker, "Fallacy of the five-year survival in lung cancer," *New England Journal of Medicine* 299: 1397–1401 (1978).

McNeil, Barbara J., Ralph Weichselbaum, and Stephen G. Pauker, "Speech and survival: Tradeoffs between quality and quantity of life in laryngeal cancer," *New England Journal of Medicine* 305: 982–987 (1981).

McNeil, Barbara J., Stephen G. Pauker, Harold C. Sox, Jr., and Amos Tversky, "On the elicitation of preferences for alternative therapies," *New England Journal of Medicine* 306: 1259–1262 (1982).

McNeill, William H., *The Pursuit of Power: Technology, Armed Force, and Society Since A.D. 1000* (The University of Chicago Press, Chicago, 1982).

McRae, Ginger, "A critical overview of U.S. acupuncture regulation," *Journal of Health Politics, Policy and Law* 7: 163–196 (1982).

Meador, Roy, *Franklin—Revolutionary Scientist* (Ann Arbor Science Publishers, Ann Arbor, Mich., 1975).

Medawar, Peter Brian, *Induction and Intuition in Scientific Thought* (Memoirs of the American Philosophical Society, volume 75, American Philosophical Society, Philadelphia, 1969).

Medawar, Peter Brian, *The Hope of Progress* (Methuen, London, 1972).

Melosi, Martin V., ed., *Pollution and Reform in American Cities, 1870–1920* (University of Texas Press, Austin, 1980).

Menzel, Paul T., *Medical Costs, Moral Choices: A Philosophy of Health Care Economics in America* (Yale University Press, New Haven, Conn., 1983).

Merkle, Judith A., *Management and Ideology, the Legacy of the International Scientific Management Movement* (University of California Press, Berkeley, Calif., 1980).

Merrill, Richard A., and Peter Barton Hutt, *Food and Drug Law* (The Foundation Press, Mineola, N.Y., 1980).

Merton, Robert K., "Science and technology in a democratic order," *Journal of Legal and Political Sociology* 1: 115–126 (1942); reprinted as "The normative structure of science," 267–278 of Robert K. Merton, *The Sociology of Science* (The University of Chicago Press, Chicago, 1973).

Metzger, Walter P., "Academic freedom and scientific freedom," *Daedalus* 107, no. 2: 93–114 (1978).

Michalos, Alex C., "A reconsideration of the idea of a science court," 10–28 of Paul T. Durbin, ed., *Research in Philosophy and Technology* 3 (Jai Press, Greenwich, Conn., 1980).

Mingle, J. O., and C. E. Reagan, "Legal and moral responsibilities of the engineer," *Chemical Engineering Progress* 76, no. 12: 15– 23 (1980).

Miser, Hugh J., "Operations research and systems analysis," *Science* 209: 139–146 (1980).

Mitcham, Carl, "Types of technology," 229–294 of Paul T. Durbin, ed., *Research in Philosophy and Technology* 1 (Jai Press, Greenwich, Conn., 1978).

Montagu, Ashley, ed., *Sociobiology Examined* (Oxford University Press, New York, 1980).

Mooney, Gavin H., *The Valuation of Human Life* (Macmillan, London, 1977).

Morgan, Joan, and W. J. Whelan, eds., *Recombinant DNA and Genetic Experimentation* (Pergamon Press, New York, 1979).

Morgan, M. Granger, William R. Fish, Samuel C. Morris, and Alan K. Meier, "Sulfur control in coal fired power plants: A probabilistic approach to policy analysis," *Journal of the Air Pollution Control Association* 28: 993–997 (1978a).

Morgan, M. Granger, Samuel C. Morris, Alan K. Meier, and Debra L. Shenk, "A probabilistic methodology for estimating air pollution health effects from coal-fired power plants," *Energy Systems and Policy* 2: 287–310 (1978b).

Morgan, M. Granger, Deborah A. L. Amaral, Max Henrion, and Samuel C. Morris, *Technological Uncertainty in Policy Analysis* (offprint, Department of Engineering and Public Policy, Carnegie–Mellon University, Pittsburgh, 1982).

Morris, Richard Knowles, and Michael W. Fox, eds., *On the Fifth Day: Animal Rights & Human Ethics* (Acropolis Books, Washington, D.C., 1978).

Moss, Thomas H., and David L. Sills, eds., "The Three Mile Island nuclear accident: Lessons and implications," *Annals of the New York Academy of Sciences* 365: 1–343 (1981).

Mouat, Frederic J., "History of the Statistical Society of London," *Journal of the Statistical Society* 50: 14–59 (1885).

Mulkay, M. J., "Sociology of the scientific research community," 93–148 of Ina Spiegel-Rösing and Derek de Solla Price, eds., *Science, Technology and Society: A Cross-Disciplinary Perspective* (Sage Publications, London and Beverly Hills, Calif., 1977).

Mulkay, Michael, *Science and the Sociology of Knowledge* (George Allen & Unwin, London, 1979).

Murray, Thomas H., and Arthur L. Caplan, eds., *Which Babies Shall Live? Humanistic Implications of the Care of Imperiled Newborns* (Humana Press, Clifton, N.J., 1985).

Mynatt, F. R., "Nuclear reactor safety research since Three Mile Island," *Science* 216: 131–135 (1982).

Myrdal, Gunnar, *Value in Social Theory* (Harper & Brothers, New York, 1958).

Myrdal, Gunnar, *Objectivity in Social Research* (Pantheon Books, New York, 1969).

Nagel, Ernest, *The Structure of Science: Problems in the Logic of Scientific Explanation* (Harcourt, Brace, and World, New York, 1961).

Nash, Leonard K., *The Nature of the Natural Sciences* (Little, Brown, Boston, 1963).

National Academy of Sciences, Committee on Literature Survey of Risks Associated with Nuclear Power, of the Committee on Science and Public Policy, *Risks Associated with Nuclear Power: A Critical Review of the Literature* (National Academy of Sciences, Washington, D.C., 1979).

National Academy of Sciences/National Research Council, Advisory Committee on the Biological Effects of Ionizing Radiations, *The Effects on Populations of Exposure to Low Levels of Ionizing Radiation* (NAS/NRC, Washington, D.C., 1972).

National Academy of Sciences/National Research Council, Advisory Committee on the Biological Effects of Ionizing Radiation, *Considerations of Health Benefit–Cost Analysis for Activities Involving Ionizing Radiation Exposure and Alternatives* (National Academy of Sciences, Washington, D.C., 1977).

National Academy of Sciences/National Research Council, Board on International Scientific Exchange, *Review of U.S.–U.S.S.R. Interacademy Exchanges and Relations* (National Academy of Sciences, Washington, D.C., 1977).

National Academy of Sciences/National Research Council, Climatic Impact Committee, *Environmental Impact of Stratospheric Flight* (National Academy of Sciences, Washington, D.C., 1975).

National Academy of Sciences/National Research Council, Committee on Chemistry and Physics of Ozone Depletion and the Committee on Biological Effects of Increased Solar Ultraviolet Radiation, *Causes and Effects of Stratospheric Ozone Reduction: An Update* (National Academy Press, Washington, D.C., 1982).

National Academy of Sciences/National Research Council, Committee on Impacts of Climatic Change, *Halocarbons: Environmental Effects of Chlorofluoromethane Release* (National Academy of Sciences, Washington, D.C., 1976).

National Academy of Sciences/National Research Council, Committee on Nitrite and Alternative Curing Agents in Food, *The Health Effects of Nitrate, Nitrite, and N-Nitroso Compounds* (National Academy Press, Washington, D.C., 1981).

National Academy of Sciences/National Research Council, Committee on Nuclear and Alternative Energy Systems, *Energy in Transition, 1985–2010* (W. H. Freeman, San Francisco, 1980).

National Academy of Sciences/National Research Council, Committee on Prototype Explicit Analyses for Pesticides, *Regulating Pesticides* (National Academy of Sciences, Washington, D.C., 1980).

National Academy of Sciences/National Research Council, Food and Nutrition Board, *Toward Healthful Diets* (National Academy of Sciences, Washington, D.C., 1980).

National Academy of Sciences/National Research Council, Panel on the Public Policy Implications of Earthquake Prediction, *Earthquake Prediction and Public Policy* (National Academy of Sciences, Washington, D.C., 1975).

National Academy of Sciences/National Research Council, Pesticide Information Review and Evaluation Committee, *An Evaluation of the Carcinogenicity of Chlordane and Heptachlor* (National Academy of Sciences, Washington, D.C., October 1977).

National Academy of Sciences/National Research Council, Subcommittee on the National Halothane Study, *The National Halothane Study* (Washington, D.C., 1969).

National Academy of Sciences, Panel on Scientific Communication and National Security, of the Committee on Science, Engineering, and Public Policy, *Scientific Communication and National Security* (National Academy Press, Washington, D.C., 1982).

National Research Council, Committee on Basic Research in the Behavioral and Social Sciences, *Behavioral and Social Science Research: A National Resource* (National Academy Press, Washington, D.C., 1982).

National Research Council, Committee on Health Care Resources in the Veterans Administration, *Health Care for American Veterans* (National Academy of Sciences, Washington, D.C., 1977).

National Research Council, Committee on Planetary Biology and Chemical Evolution, *Recommendations on Quarantine Policy for Mars, Jupiter, Saturn, Uranus, Neptune, and Titan* (National Academy of Sciences, Washington, D.C., 1978).

National Research Council, Committee on Substance Abuse and Habitual Behavior, *An Analysis of Marijuana Policy* (National Academy Press, Washington, D.C., 1982).

National Research Council, Committee on the Institutional Means for Assessment of Risks to Public Health, *Risk Assessment in the Federal Government: Managing the Process* (National Academy Press, Washington, D.C., 1983).

National Research Council, Committee on Underground Coal Mine Safety, *Toward Safer Underground Coal Mines* (National Academy Press, Washington, D.C., 1982).

National Research Council, Diesel Impacts Study Committee, *Diesel Cars: Benefits, Risks, and Public Policy* (National Academy Press, Washington, D.C., 1982).

National Research Council, "Enrichment of flour and bread: A history of the movement," *Bulletin of the National Research Council*, no. 110 (National Research Council, Washington, D.C., 1944).

National Research Council, Panel on Groundwater Contamination, *Groundwater Contamination* (National Academy Press, Washington, D.C., 1984).

National Research Council, Panel on Survey Measurements of Subjective Phenomena, *Surveys of Subjective Phenomena* (National Academy Press, Washington, D.C., 1981).

Neustadt, Richard E., and Harvey V. Fineberg, *The Epidemic that Never Was: Policy-Making and the Swine Flu Affair* (Vintage Publishers, New York, 1983).

Newhouse, John, *The Sporty Game* (Alfred A. Knopf, New York, 1982).

Nichols, Rodney W., "Some practical problems of scientist-advisors," *Minerva* 10: 603–613 (1972).

Nisbet, Robert, "Knowledge dethroned," *New York Times Magazine*: 34ff. (September 28, 1975).

Noble, David F., *America by Design: Science, Technology, and the Rise of Corporate Capitalism* (Alfred A. Knopf, New York, 1977).

Nozick, Robert, *Anarchy, State, and Utopia* (Basic Books, New York, 1974).

Nuclear Energy Policy Study Group, *Nuclear Power Issues and Choices* (Ballinger, Cambridge, Mass., 1977).

O'Brien, David M., and Donald A. Marchand, eds., *The Politics of Technology Assessment* (Lexington Books, Lexington, Mass., 1982).

Ogburn, William Fielding, *The Social Effects of Aviation* (Houghton Mifflin, Boston, 1946).

Okrent, David, *Nuclear Reactor Safety* (University of Wisconsin Press, Madison, Wis., 1981).

Okrent, David, and Dade Moeller, "Implications for reactor safety of the accident at Three Mile Island, Unit 2," *Annual Review of Energy* 6: 43–88 (1981).

Omenn, Gilbert S., "Predictive identification of hypersusceptible individuals," *Journal of Occupational Medicine* 24: 369–374 (1982).

Oppenheimer, J. Robert, "Physics in the contemporary world," *Bulletin of the Atomic Scientists* 4, no. 3: 65–68 (1948).

Organisation for Economic Co-operation and Development, *Technology on Trial: Public Participation in Decision-making Related to Science and Technology* (OECD, Paris, 1979).

Outhwaite, William, *Concept Formation in Social Science* (Routledge & Kegan Paul, London, 1983).

Papineau, David, *For Science in the Social Sciences* (Macmillan, London, 1978).

Parekh, Bhikhu C., *Bentham's Political Thought* (Croom Helm, London, 1973).

Parker, Justice Roger Jocelyn, *The Windscale Inquiry*, report presented to the Secretary of State for the Environment (Her Majesty's Stationery Office, London, 1978).

Parsons, Talcott, *The Social System* (Free Press, Glencoe, Ill., 1951).

Passmore, John, *Man's Responsibility for Nature: Ecological Problems and Western Traditions* (Charles Scribner's Sons, New York, 1974).

Paul, Jeffrey, *Reading Nozick: Essays on Anarchy, State, and Utopia* (Rowman and Littlefield, Totowa, N.J., 1981).

Paul, John R., *A History of Poliomyelitis* (Yale University Press, New Haven, Conn., 1971).

Pechman, Joseph A., and P. Michael Timpane, eds., *Work Incentives and Income Guarantees: The New Jersey Negative Income Tax Experiment* (The Brookings Institution, Washington, D.C., 1975).

Pellegrino, Edmund D., *Humanism and the Physician* (University of Tennessee Press, Knoxville, Tenn., 1979).

Pellegrino, Edmund D., and David C. Thomasma, *A Philosophical Basis of Medical Practice* (Oxford University Press, New York, 1981).

Pernick, Martin S., *A Calculus of Suffering: Pain, Professionalism, and Anesthesia in 19th Century America* (Columbia University Press, New York, 1985).

Perrucci, Robert, and Joel E. Gerstl, *Profession Without Community: Engineers in American Society* (Random House, New York, 1969).

Phibbs, Ciaran S., Ronald L. Williams, and Roderic H. Phibbs, "Newborn risk factors and costs of neonatal intensive care," *Pediatrics* 68: 313–321 (1981).

Phillips, Leon, A., "Comparative evaluation of the effect of a high yield criteria list upon skull radiography," *Journal of the American College of Emergency Physicians* 8: 106–109 (1979).

Piel, Gerard, "Inquiring into inquiry," *Hastings Center Report* 6, no. 4: 18–19 (1976).

Polanyi, Michael, *Personal Knowledge*, corrected ed. (The University of Chicago Press, Chicago, 1962a).

Polanyi, Michael, "The republic of science: Its political and economic theory," *Minerva* 1: 54–73 (1962b).

Population Reference Bureau, Inc., "U.S. Population: Where we are; where we're going," *Population Bulletin* 37, no. 2 (1982).

Potter, Harry R., and Heather J. Norville, "Social values inherent in policy statements: An evaluation of an energy technology assessment," 177–189 of Dean E. Mann, ed., *Environmental Policy Formation* (Lexington Books, Lexington, Mass., 1981).

Prewitt, Kenneth, "Early warning systems," *Society* 18, no. 6: 3–23 (1981).

Price, Don K., *The Scientific Estate* (Harvard University Press, Cambridge, Mass., 1965).

Pye, Veronica I., Ruth Patrick, and John Quarles, *Groundwater Contamination in the United States* (University of Pennsylvania Press, Philadelphia, 1983).

Rabi, I. I., "Scientists and social responsibility: Publication is the chief responsibility," *Bulletin of the Atomic Scientists* 4, no. 3: 73 (1948).

Rabinow, Paul, and William M. Sullivan, eds., *Interpretive Social Science: A Reader* (University of California Press, Berkeley, Calif., 1979).

Rae, Douglas, *Equalities* (Harvard University Press, Cambridge, Mass., 1981).

Raleigh, C. B., "Scientists' responsibility for public information," editorial, *Science* 213: 499 (1981).

Ramo, Simon, "Regulation of technological activities: A new approach," *Science* 213: 837–842 (1981).

Ramsey, Paul, *Ethics at the Edges of Life: Medical and Legal Intersections* (Yale University Press, New Haven, Conn., 1978).

Rapoport, Anatol, "Scientific approach to ethics," *Science* 125: 796–799 (1957).

Ravetz, Jerome, *Scientific Knowledge and its Social Problems* (Oxford University Press, Oxford, 1971).

Rawls, John, *A Theory of Justice* (Harvard University Press, Cambridge, Mass., 1971).

Rawls, John, "Kantian constructivism in moral theory," *Journal of Philosophy* 77: 515–572 (1980).

Rawls, John, "Social unity and primary goods," 159–185 of Amartya Sen and Bernard Williams, eds., *Utilitarianism and Beyond* (Cambridge University Press, Cambridge, 1982).

Read, J. Leighton, Robert J. Quinn, Donald M. Berwick, Harvey V. Fineberg, and Milton C. Weinstein, "Preferences for health outcomes: Comparison of assessment methods," *Medical Decision Making* 4: 315–329 (1984).

Regan, Tom, *All That Dwell Therein: Animal Rights and Environmental Ethics* (University of California Press, Berkeley, Calif., 1982).

Regan, Tom, *The Case for Animal Rights* (University of California Press, Berkeley, Calif., 1983).

Regan, Tom, and Peter Singer, eds., *Animal Rights and Human Obligations* (Prentice-Hall, Englewood Cliffs, N.J., 1976).

Reich, Walter, "Psychiatric diagnosis as an ethical problem," 61–88 of Sidney Bloch and Paul Chodoff, eds., *Psychiatric Ethics* (Oxford University Press, New York, 1981).

Reid, Robert W., *Tongues of Conscience: War and the Scientist's Dilemma* (Archibald Constable, London, 1969).

Relman, Arnold S., "The new medical-industrial complex," *New England Journal of Medicine* 303: 963–970 (1980).

Rescher, Nicholas, "What is value change? A framework for research," 68–98 of Kurt Baier and Nicholas Rescher, eds., *Values and the Future* (Free Press, New York, 1969).

Reynolds, Paul Davidson, *Ethical Dilemmas and Social Science Research* (Jossey-Bass, San Francisco, 1980).

Rhoads, Steven E., "How much should we spend to save a life?" *Public Interest*, no. 51: 92 (1978); reprinted in expanded form as 285–311 of Steven E. Rhoads, ed., *Valuing Life: Public Policy Dilemmas* (Westview Press, Boulder, Colo., 1980).

Rhoads, Steven E., ed., *Valuing Life: Public Policy Dilemmas* (Westview Press, Boulder, Colo., 1980).

Richter, Jean Paul, ed., *The Notebooks of Leonardo da Vinci* (Dover, New York, 1970).

Rickover, Hyman G., "A humanistic technology," *Nature* 208: 721–726 (1965).

Rivlin, Alice M., "Forensic social science," *Harvard Educational Review* 43: 61–75 (1975).

Rivlin, Alice M., and P. Michael Timpane, eds., *Ethical and Legal Issues of Social Experimentation* (The Brookings Institution, Washington, D.C., 1975).

Roberts, Marc J., "On the nature and condition of social science," *Daedalus* 103, no. 4: 47–64 (1974).

Robertson, John A., "The scientist's right to research: A constitutional analysis," 51 *Southern California Law Review*: 1203–1279 (1978).

Robinson, Joan, *Economic Philosophy* (Aldine, Chicago, 1963).

Robitscher, Jonas, *The Powers of Psychiatry* (Houghton Mifflin, Boston, 1980).

Rogers, Michael, *Biohazard* (Alfred A. Knopf, New York, 1977).

Rokeach, Milton, *The Nature of Human Values* (Free Press, New York, 1973).

Rolston, Holmes, III, "Are values in nature subjective or objective?" *Environmental Ethics* 4: 125–151 (1982).

Rosen, Laurence, "The anthropologist as expert witness," *American Anthropologist* 79: 555–577 (1977).

Rosenberg, Nathan, *Inside the Black Box: Technology and Economics* (Cambridge University Press, Cambridge, 1982).

Rosner, Fred, and J. David Bleich, eds., *Jewish Bioethics* (Sanhedrin Press, New York, 1979).

Ross, H. Laurence, *Deterring the Drinking Driver* (Lexington Books, Lexington, Mass., 1982).

Rothschild, Joan, *Machina Ex Dea: Feminist Perspectives on Technology* (Pergamon Press, New York, 1983).

Rowan, Andrew N., *Of Mice, Models, and Men: A Critical Evaluation of Animal Research* (State University of New York Press, New York, 1984).

Royal Society Study Group on Risk, *Risk Assessment* (The Royal Society, 6 Carlton House Terrace, London SWIY 5AG, 1983).

Ruckelshaus, William D., "Risk in a free society," *Risk Analysis* 4: 157–162 (1984).

Rueth, Nancy, "Ethics and the boiler code," *Mechanical Engineering* 97: 34–36 (1975).

Runciman, W. G., *A Treatise on Social Theory. Volume I: The Methodology of Social Theory* (Cambridge University Press, Cambridge, 1983).

Russell, Bertrand, *Sceptical Essays* (W. W. Norton, New York, 1928).

Russell, Bertrand, "The social responsibilities of scientists," *Science* 131: 391–392 (1960).

Sabia, Daniel R., Jr., and Jerald Wallulis, eds., *Changing Social Science: Critical Theory and Other Critical Perspectives* (State University of New York Press, Albany, N.Y., 1983).

Salem, Steven L., Kenneth A. Solomon, and Michael S. Yesley, *Issues and problems in inferring a level of acceptable risk* (Rand Corporation no. R-2561-DOE, Santa Monica, Calif. 90406, 1980).

Sandel, Michael J., *Liberalism and the Limits of Justice* (Cambridge University Press, Cambridge, 1982).

Sapolsky, Harvey M., "Science, technology and military policy," 443–472 of Ina Spiegel-Rösing and Derek de Solla Price, eds., *Science, Technology and Society* (Sage Publications, London and Beverly Hills, Calif., 1977).

Sartorius, Rolf, ed., *Paternalism* (University of Minnesota Press, Minneapolis, Minn., 1983).

Schattschneider, Elmer E., *The Semisovereign People* (Dryden Press, Hinsdale, Ill., 1960).

Schuler, Heinz, *Ethical Problems in Psychological Research*, translated by Margaret S. Woodruff and Robert A. Wicklund (Academic Press, New York, 1982).

Schultz, Stanley K., and Clay McShane, "To engineer the metropolis: Sewers, sanitation, and city planning in late-nineteenth-century America," *Journal of American History* 65: 389–411 (1978).

Seeley, John R., "Social science? Some probative problems," 53–65 of Maurice Stein and Arthur Vidich, eds., *Sociology on Trial* (Prentice-Hall, Englewood Cliffs, N.J., 1963).

Seiden, Anne M., "Overview: Research on the psychiatry of women. II. Women in families, work, and psychotherapy," *American Journal of Psychiatry* 133: 1111–1123 (1976).

Select Committee on GRAS Substances, "Insights on food safety evaluation," NTIS no. PB83-154146 (National Technical Information Service, Springfield, Va., 1982).

Sen, Amartya, "Accounts, actions and values: Objectivity of social science," 87–107 of Christopher Lloyd, ed., *Social Theory and Political Practice* (Oxford University Press, Oxford, 1983).

Sen, Amartya, and Bernard Williams, eds., *Utilitarianism and Beyond* (Cambridge University Press, Cambridge, 1982).

Shaw, Margery W., letter to the editor, *Science* 209: 751–752 (1980).

Shelp, Earl E., ed., *Beneficence and Health Care* (D. Reidel, Boston, 1982).

Shemilt, L. W., ed., *Chemistry and World Food Supplies: The New Frontiers* (Pergamon Press, New York, 1983).

Shils, Edward, *The Calling of Sociology* (The University of Chicago Press, Chicago, 1980).

Shils, Edward, *Tradition* (The University of Chicago Press, Chicago, 1981).

Shryock, Richard H., "The history of quantification in medical science," *Isis* 52: 215–237 (1961).

Siegler, Mark, "Confidentiality in medicine—a decrepit concept," *New England Journal of Medicine* 307: 1518–1521 (1982).

Sievers, Bruce, "Believing in social science: The ethics and epistemology of public opinion research," 320–342 of Norma Haan, Robert N. Bellah, Paul Rabinow, and William M. Sullivan, eds., *Social Science As Moral Inquiry* (Columbia University Press, New York, 1983).

Sills, David L., C. P. Wolf, and Vivien B. Shelanski, eds., *Accident at Three Mile Island: The Human Dimensions* (Westview Press, Boulder, Colo., 1982).

Silverstein, Arthur M., *Pure Politics and Impure Science: The Swine Flu Affair* (The Johns Hopkins University Press, Baltimore, 1981).

Simon, Herbert A., "The behavioral and social sciences," *Science* 209: 72–78 (1980).

Simon, Julian L., "Resources, population, environment: An oversupply of false bad news," *Science* 208: 1431–1437 (1980).

Simon, Julian L., and Herman Kahn, eds., *The Resourceful Earth: A Response to Global 2000* (Basil Blackwell, London and New York, 1984).

Sinclair, Bruce, *Early Research at the Franklin Institute: The Investigation into the Causes of Steam Boiler Explosions: 1830–1837* (The Franklin Institute, Philadelphia, 1966).

Sinclair, Bruce, *A Centennial History of the American Society of Mechanical Engineers* (University of Toronto Press, Toronto, 1980).

Sinden, John A., and Albert C. Worrell, *Unpriced Values: Decisions without Market Prices* (John Wiley & Sons, New York, 1979).

Singer, Peter, "Ten years of animal liberation," *New York Review of Books* 31, no. 21/22: 46–52 (1985).

Sinsheimer, Robert L., "The presumptions of science," *Daedalus* 107, no. 2: 23–35 (1978).

Small, Albion W., *The Meaning of Social Science* (The University of Chicago Press, Chicago, 1910).

Smith, Alice Kimball, *A Peril and a Hope: The Scientists' Movement in America, 1945–1947* (The University of Chicago Press, Chicago, 1965).

Smith, Cyril Stanley, "A historical view of one area of applied science—metallurgy," 57–71 of National Academy of Sciences, Panel on Applied Science and Technological Progress, *Applied Science and Technological Progress* (U.S. Government Printing Office, Washington, D.C., June 1967).

Social Science Research Council, "Federal funding for the social sciences: Threats and responses," *Social Science Research Council Items* 35: 33–47 (1981).

Society of Toxicology, ED_{01} Task Force, "Re-examination of the ED_{01} Study," *Fundamental and Applied Toxicology* 1: 25–128 (1981).

51 *Southern California Law Review*, 969–1554, special issue on "Biotechnology and the law: Recombinant DNA and the control of scientific research" (1978).

Spero, Moshe Halevi, "Modern psychotherapy and halakhic values: An approach toward consensus in values and practice," *Journal of Medicine and Philosophy* 8: 287–316 (1983).

Sperry, Roger, *Science and Moral Priority: Merging Mind, Brain, and Human Values* (Columbia University Press, New York, 1983).

Spritzer, Robert L., Janet B. W. Forman, and John Nee, "DSM-III Field trials: Initial interrater diagnostic reliability," *American Journal of Psychiatry* 136: 815–817 (1979).

Stanley, Manfred, "The mystery of the commons: On the indispensability of civic rhetoric," *Social Research* 50: 851–883 (1984).

Starr, Paul, *The Social Transformation of American Medicine* (Basic Books, New York, 1982).

Starr, Paul, and Ross Corson, "Who will have the numbers? The rise of the statistical services industry and the politics of public data," in William Alonso and Paul Starr, eds., *The Political Economy of National Statistics* (Russell Sage Foundation, New York, in press).

Stent, Gunther S., ed., *Morality as a Biological Phenomenon: The Presuppositions of Sociobiological Research*, 2nd ed. (University of California Press, Berkeley, Calif., 1980).

Stetten, DeWitt, Jr., "Valedictory by the chairman of the NIH Recombinant DNA Molecular Program Advisory Committee," *Gene* 3: 265–268 (1978).

Stigler, George J., "Economics or ethics?" 143–191 of Sterling M. McMurrin, ed., *The Tanner Lectures on Human Values*, II (University of Utah Press, Salt Lake City, Utah, 1981).

Stokey, Edith, and Richard Zeckhauser, *A Primer for Policy Analysis* (W. W. Norton, New York, 1978).

Stone, Alan A., *Law, Psychiatry, and Morality* (American Psychiatric Press, Washington, D.C., 1984).

Stone, Christopher D., *Should Trees Have Standing? Toward Legal Rights for Natural Objects* (William Kaufmann, Los Altos, Calif., 1974).

Stone, Deborah A., "Physicians as gatekeepers: Illness certification as a rationing device," *Public Policy* 27: 227–254 (1979).

St.-Pierre, L. E., ed., *Resources of Organic Matter for the Future* (Multiscience Publications, Suite 175, 1253 McGill College, Montreal, Quebec H3B 2Y5, 1978).

Sun, Marjorie, "Lawyers flush out toxic shock data," *Science* 224: 132–134 (1984).

Suppe, Frederick, ed., *The Structure of Scientific Theories*, 2nd ed. (University of Illinois Press, Urbana, Ill., 1977).

Swan, Rita, "Faith healing, Christian Science, and the medical care of children," *New England Journal of Medicine* 309: 1639–1641 (1983).

Swartzman, Daniel, Richard A. Liroff, and Kevin G. Croke, eds., *Cost–Benefit Analysis and Environmental Regulation* (The Conservation Foundation, 1717 Massachusetts Ave. NW, Washington, D.C. 20036, 1982).

Swazey, Judith P., James R. Sorenson, and Cynthia B. Wong, "Risks and benefits, rights and responsibilities: A history of the recombinant DNA research controversy," 51 *Southern California Law Review*: 1019–1078 (1978).

Talbot, Allan R., *Settling Things: Six Cases in Environmental Mediation* (The Conservation Foundation, 1717 Massachusetts Ave. NW, Washington, D.C. 20036, 1983).

Talbot, Nathan A., "The position of the Christian Science church," *New England Journal of Medicine* 309: 1641–1644 (1983).

Tanur, Judith M., "Advances in methods for large-scale surveys and experiments," 294–372 of National Research Council, Committee on Basic Research in the Behavioral and Social Sciences, *Behavioral and Social Science Research: A National Resource*, Part II (National Academy Press, Washington, D.C., 1982).

Task Force of Past Presidents of the Society of Toxicology, "Animal data in hazard evaluation: Paths and pitfalls," *Fundamental and Applied Toxicology* 2: 101–107 (1982).

Task Force of the Presidential Advisory Group on Anticipated Advances in Science and Technology, "The science court experiment: An interim report," *Science* 193: 653–656 (1976).

Tatum, Edward L., "Perspectives from physiological genetics," 20–34 of T. M. Sonnenborn, ed., *The Control of Human Heredity and Evolution* (Macmillan, New York, 1965).

Taylor, Charles, "Neutrality in political science," 25–57 of Peter Laslett and W. G. Runciman, eds., *Philosophy, Politics and Society*, 3rd series (Barnes & Noble, New York, 1967).

Teller, Edward, "I was the only victim of Three-Mile Island," advertisement, *Wall Street Journal* (July 31, 1979).

Teller, Edward, with Allen Brown, *The Legacy of Hiroshima* (Doubleday, Garden City, N.Y., 1962).

Thomas, Lewis, "The hazards of science," *New England Journal of Medicine* 296: 324–328 (1977).

Thomas, William L., Jr., ed., *Man's Role in Changing the Face of the Earth* (The University of Chicago Press, Chicago, Ill., 1956).

Thompson, Dennis, "Ascribing responsibility to advisers in government," *Ethics* 93: 546–560 (1983).

Thomson, J. J., *The Corpuscular Theory of Matter* (Archibald Constable, London, 1907).

Thurow, Lester C., *Dangerous Currents: The State of Economics* (Random House, New York, 1983).

Tobin, Richard J., *The Social Gamble: Determining Acceptable Levels of Air Quality* (Lexington Books, Lexington, Mass., 1979).

Tocqueville, Alexis de, *Democracy in America* [1835], translated by George Lawrence, edited by J. P. Mayer and Max Lerner (Harper & Row, New York, 1966).

Toulmin, Stephen, *Human Understanding* (Princeton University Press, Princeton, N.J., 1972).

Toulmin, Stephen, "From form to function: Philosophy and history of sciences in the 1950s and now," *Daedalus* 106, no. 3: 143–162 (1977).

Trescott, Martha Moore, "Lillian Moller Gilbreth and the founding of modern industrial engineering," 23–37 of Joan Rothschild, ed., *Machina Ex Dea* (Pergamon Press, New York, 1983).

Trilling, Lionel, "Sex and science: The Kinsey Report," *Partisan Review* 15: 460–476 (1948); reprinted as 210–225 of Lionel Trilling, *The Liberal Imagination* (Harcourt Brace Jovanovich, New York, 1979).

Unger, Stephen H., *Controlling Technology: Ethics and the Responsible Engineer* (Holt, Rinehart and Winston, New York, 1982).

Union of Concerned Scientists, "The risks of nuclear power reactors: A review of the NRC Reactor Safety Study WASH-1400 (NUREG-75/014)" (UCS, 26 Church St., Cambridge, Mass. 02238, 1977).

Union of Concerned Scientists, *The Fallacy of Star Wars: Why Space Weapons Won't Work* (Random House, New York, 1984).

U.S. Army, Adjutant General's Department, *Trials of War Criminals Before Nuernberg Military Tribunals, under Control Council Law No. 10 (October 1946–April 1949)*, vol. 2 (U.S. Government Printing Office, Washington, D.C., 1947).

U.S. Atomic Energy Commission, "In the matter of J. Robert Oppenheimer," transcript of hearing before Personnel Security Board April 12–May 6, 1954 (U.S. Government Printing Office, Washington, D.C., 1954).

U.S. Congress, Congressional Research Service, *The Evolution and Dynamics of National Goals in the United States*, prepared for the Committee on Interior and Insular Affairs, U.S. Senate, 92nd Congress, 1st Session (U.S. Government Printing Office, Washington, D.C., 1971).

U.S. Congress, Congressional Research Service, *Genetic Engineering, Human Genetics, and Cell Biology: Evolution of Technological Issues*, Supplemental Report II, prepared for the Subcommittee on Science, Research and Technology, of the Committee on Science and Technology, U.S. House of Representatives, 94th Congress, 2nd Session (U.S. Government Printing Office, Washington, D.C., December 1976).

U.S. Congress, Congressional Research Service, *Science, Technology, and American Diplomacy*, prepared for the Committee on International Relations, of the U.S. House of Representatives (U.S. Government Printing Office, Washington, D.C., 1977).

U.S. Congress, Congressional Research Service, *Technical Information for Congress*, 3rd edition, prepared for the Subcommittee on Science, Research and Technology, of the Committee on Science and Technology, U.S. House of Representatives (U.S. Government Printing Office, Washington, D.C., July 1979).

U.S. Congress, Office of Technology Assessment, *Assessing the Efficacy and Safety of Medical Technologies* (U.S. Government Printing Office, Washington, D.C., 1978).

U.S. Congress, Office of Technology Assessment, *The Implications of Cost-Effectiveness Analysis on Medical Technology* (U.S. Government Printing Office, Washington, D.C., August 1980).

U.S. Congress, Office of Technology Assessment, "The efficacy and cost-effectiveness of psychotherapy," Background Paper no. 3 to *The Implications of Cost-Effectiveness Analysis of Medical Technology* (U.S. Government Printing Office, Washington, D.C., October 1980).

U.S. Congress, Office of Technology Assessment, *Impacts of Applied Genetics: Microorganisms, Plants, and Animals* (U.S. Government Printing Office, Washington, D.C., April 1981).

U.S. Congress, Office of Technology Assessment, *Assessment of Technologies for Determining Cancer Risks from the Environment* (U.S. Government Printing Office, Washington, D.C., June 1981).

U.S. Congress, Office of Technology Assessment, *Technology and Handicapped People* (U.S. Government Printing Office, Washington, D.C., May 1982).

U.S. Congress, Office of Technology Assessment, *The Role of Genetic Testing in the Prevention of Occupational Disease* (U.S. Government Printing Office, Washington, D.C., April 1983).

U.S. Congress, Office of Technology Assessment, *Protecting the Nation's Groundwater from Contamination* (U.S. Government Printing Office, Washington, D.C., 1984).

U.S. Council on Environmental Quality and Department of State, *The Global 2000 Report to the President* (U.S. Government Printing Office, Washington, D.C., 1980; reprinted by Penguin Books, New York, 1982).

U.S. Department of Defense, Defense Science Board, Task Force on Export of U.S. Technology, *An Analysis of Export Control of U.S. Technology: A DoD Perspective* (U.S. Department of Defense, Washington, D.C., 1976).

U.S. Department of Health and Human Services, Office of the Secretary, "Final regulations amending basic HHS policy for the protection of human research subjects: Final rule," 46 *Federal Register*: 8366–8391 (January 26, 1981).

U.S. Department of Health, Education, and Welfare, *Report of the Secretary's Commission on Pesticides and Their Relationship to Environmental Health* (U.S. Government Printing Office, Washington, D.C., 1969).

U.S. Environmental Protection Agency, "Reactor Safety Study (WASH-1400): A review of the final report," (EPA no. 520/3-76-009, Washington, D.C., June 1976).

U.S. Environmental Protection Agency, *Ground-Water Protection Strategy for the Environmental Protection Agency* (Washington, D.C., August 1984).

U.S. Food and Drug Administration, "Good laboratory practice regulations," 43 *Federal Register*: 59,986–60,013 (December 22, 1978).

U.S. House of Representatives, Government Activities and Transportation Subcommittee, of the Committee on Government Operations, *Hearings on FAA Certification of the SST Concorde*, 94th Congress, 1st Session (July 24, 1975).

U.S. House of Representatives, Legislation and National Security Subcommittee, of the Committee on Government Operations, *Hearings on Federal Government Statistics and Statistical Policy*, 97th Congress, 2nd Session (June 3, 1982).

U.S. House of Representatives, Subcommittee on Census and Population, of the Committee on Post Office and Civil Service, *Hearings on Impact of Budget Cuts on Federal Statistical Programs*, 97th Congress, 2nd Session (March 16, 1982).

U.S. House of Representatives, Subcommittee on Domestic Marketing, Consumer Relations, and Nutrition, of the Committee on Agriculture, *Hearings on the National Academy of Sciences Report on Healthful Diets*, 96th Congress, 2nd Session (June 18–19, 1980).

U.S. House of Representatives, Subcommittee on Investigations and Oversight, of the Committee on Science and Technology, *Hearings on Fraud in Biomedical Research*, 97th Congress, 1st Session (March 31 and April 1, 1981).

U.S. House of Representatives, Subcommittee on Investigations and Oversight, of the Committee on Science and Technology, *Hearings on Genetic Screening of Workers*, 97th Congress, 2nd Session (June 22 and October 6, 1982).

U.S. House of Representatives, Subcommittee on Science, Research and Technology, of the Committee on Science and Technology, *Hearings on Science Policy Implications of DNA Recombinant Molecule Research*, 95th Congress, 1st Session (March 29,30,31; April 27,28; May 3,4,5,25,26; September 7,8, 1977).

U.S. House of Representatives, Subcommittee on Science, Research and Technology, of the Committee on Science and Technology, *Hearings on Comparative Risk Assessment*, 96th Congress, 2nd Session (May 14–15, 1980).

U.S. House of Representatives, Subcommittee on Science, Research and Technology, of the Committee on Science and Technology, *Hearings on the 1982 National Science Foundation Authorization* [525–682 cover the social and behavioral sciences budget], 97th Congress, 1st Session (January 28; March 3,5,10,12,17, 1981a).

U.S. House of Representatives, Subcommittee on Science, Research and Technology, of the Committee on Science and Technology, *Hearings on Use of Animals in Medical Research and Testing*, 97th Congress, 1st Session (October 13–14, 1981b).

U.S. National Center for Health Care Technology, "Coronary artery bypass surgery," *Journal of the American Medical Association* 246: 1645–1649 (1981).

U.S. National Center for Health Statistics, *Hearing Ability of Persons by Sociodemographic and Health Characteristics: United States* (U.S. Department of Health and Human Services no. (PHS)82-1568, Hyattsville, Md., 1982).

U.S. National Commission for the Protection of Human Subjects of Biomedical and Behavioral Research, *Research Involving Prisoners* (U.S. Government Printing Office, Washington, D.C., 1976).

U.S. National Commission for the Protection of Human Subjects of Biomedical and Behavioral Research, *Research on the Fetus* (U.S. Government Printing Office, Washington, D.C., 1976).

U.S. National Commission for the Protection of Human Subjects of Biomedical and Behavioral Research, *Psychosurgery* (U.S. Government Printing Office, Washington, D.C., 1977).

U.S. National Commission for the Protection of Human Subjects of Biomedical and Behavioral Research, *Research Involving Children* (U.S. Government Printing Office, Washington, D.C., 1977).

U.S. National Commission for the Protection of Human Subjects of Biomedical and Behavioral Research, *Ethical Guidelines for the Delivery of Health Services by DHEW [Department of Health, Education, and Welfare]* (U.S. Government Printing Office, Washington, D.C., 1978).

U.S. National Commission for the Protection of Human Subjects of Biomedical and Behavioral Research, *Research Involving Those Institutionalized as Mentally Infirm* (U.S. Government Printing Office, Washington, D.C., 1978).

U.S. National Commission for the Protection of Human Subjects of Biomedical and Behavioral Research, *The Belmont Report: Ethical Principles and Guidelines for the Protection of Human Subjects of Research* (U.S. Government Printing Office, Washington, D.C., 1978).

U.S. National Commission on Electronic Funds Transfer, *Final Report on EFT in the United States* (U.S. Government Printing Office, Washington, D.C., 1977).

U.S. National Goals Research Staff, *Toward Balanced Growth: Quantity with Quality* (U.S. Government Printing Office, Washington, D.C., 1970).

U.S. National Institutes of Health, "Coronary artery bypass surgery: Scientific and clinical aspects," *New England Journal of Medicine* 304: 680–684 (1981).

U.S. National Institute of Mental Health, *Television and Behavior: Ten Years of Scientific Progress and Implications for the Eighties* (U.S. Department of Health and Human Services no. (ADM)82-1195, Washington, D.C., 1982).

U.S. Nuclear Regulatory Commission, *Reactor Safety Study: An Assessment of Accident Risks in U.S. Commercial Nuclear Power Plants* (no. WASH-1400, or NUREG/74-014, Washington, D.C., October 1975).

U.S Nuclear Regulatory Commission, *Risk Assessment Review Group Report* (no. NUREG/CR-0400, Washington, D.C., September 1978).

U.S. Nuclear Regulatory Commission, *Human Factors Evaluation of Control Room Design and Operator Performance at Three Mile Island-2* (no. NUREG/CR-1270, Washington, D.C., January 1980).

U.S. Nuclear Regulatory Commission, *Safety Goals for Nuclear Power Plant Operation* (no. NUREG-0880, revision 1, Washington, D.C., May 1983).

U.S. Nuclear Regulatory Commission, *Probabilistic Risk Assessment (PRA) Reference Document* (no. NUREG-1050, Washington, D.C., September 1984).

U.S. Occupational Safety and Health Administration, "Occupational exposure to inorganic mercury," 48 *Federal Register*: 1864–1903 (January 14, 1983).

U.S. Office of Science and Technology Policy, "Chemical carcinogens: Review of the science and its associated principles," 49 *Federal Register*: 21,593–21,661 (May 22, 1984).

U.S. Office of the Assistant Secretary for Health and Surgeon General, *Healthy People* (U.S. Government Printing Office, Washington, D.C., 1979).

U.S. Office of the Assistant Secretary for Health and Surgeon General, *Promoting Health/Preventing Disease: Objectives for the Nation* (U.S. Government Printing Office, Washington, D.C., Fall 1980).

U.S. Office of the Assistant Secretary for Health and Surgeon General, "Promoting Health/Preventing Disease: Implementation plans for attaining the Objectives for the Nation," supplement to the September–October 1983 issue of

Public Health Reports (U.S. Government Printing Office, Washington, D.C., 1983).

U.S. President's Commission for the Study of Ethical Problems in Medicine and Biomedical and Behavioral Research, *Defining Death* (U.S. Government Printing Office, Washington, D.C., July 1981).

U.S. President's Commission for the Study of Ethical Problems in Medicine and Biomedical and Behavioral Research; American Association for the Advancement of Science, Committee on Scientific Freedom and Responsibility; and Medicine in the Public Interest, *Whistleblowing in Biomedical Research* (U.S. Government Printing Office, Washington, D.C., 1982).

U.S. President's Commission for the Study of Ethical Problems in Medicine and Biomedical and Behavioral Research, *Making Health Care Decisions* (U.S. Government Printing Office, Washington, D.C., October 1982).

U.S. President's Commission for the Study of Ethical Problems in Medicine and Biomedical and Behavioral Research, *Splicing Life: The Social and Ethical Issues of Genetic Engineering with Human Beings* (U.S. Government Printing Office, Washington, D.C., November 1982).

U.S. President's Commission for the Study of Ethical Problems in Medicine and Biomedical and Behavioral Research, *Screening and Counseling for Genetic Conditions* (U.S. Government Printing Office, Washington, D.C., February 1983).

U.S. President's Commission for the Study of Ethical Problems in Medicine and Biomedical and Behavioral Research, *Deciding to Forego Life-Sustaining Treatment* (U.S. Government Printing Office, Washington, D.C., March 1983a).

U.S. President's Commission for the Study of Ethical Problems in Medicine and Biomedical and Behavioral Research, *Implementing Human Research Regulations* (U.S. Government Printing Office, Washington, D.C., March 1983b).

U.S. President's Commission for the Study of Ethical Problems in Medicine and Biomedical and Behavioral Research, *Securing Access to Health Care* (U.S. Government Printing Office, Washington, D.C., March 1983c).

U.S. President's Commission for the Study of Ethical Problems in Medicine and Biomedical and Behavioral Research, *Summing Up* (U.S. Government Printing Office, Washington, D.C., March 1983d).

U.S. President's Commission on National Goals, *Goals for Americans* (Prentice-Hall, Englewood Cliffs, N.J., 1960).

U.S. President's Commission on the Accident at Three Mile Island, *The Need for Change: The Legacy of TMI* (U.S. Government Printing Office, Washington, D.C., 1979a).

U.S. President's Commission on the Accident at Three Mile Island, *Technical Staff Analysis Report on Alternative Event Sequences* (offprint, Washington, D.C., 1979b).

U.S. President's Commission on the Accident at Three Mile Island, *Technical Staff Analysis Report on WASH-1400/Reactor Safety Study* (offprint, Washington, D.C., 1979c).

U.S. President's Research Committee on Social Trends, *Recent Social Trends* (McGraw-Hill, New York, 1933).

U.S. President's Science Advisory Committee, *Seismic Improvement* (offprint, Washington, D.C., June 12, 1959).

U.S. President's Science Advisory Committee, *The World Food Problem* (U.S. Government Printing Office, Washington, D.C., 1967).

U.S. President's Science Advisory Committee, *Report on 2,4,5-T* (U.S. Government Printing Office, Washington, D.C., March 1971).

U.S. Senate, Committee on Aeronautical and Space Sciences, *Hearings on Educational Programs (NASA)*, 88th Congress, 1st Session (November 21–22, 1963).

U.S. Senate, Committee on Appropriations, *Hearing on H. J. Res. 468, Civil Supersonic Aircraft Development (SST)*, 92nd Congress, 1st Session (March 10–11, 1971).

U.S. Senate, Committee on Governmental Affairs, *Whistleblowers. Report on Federal Employees who Disclose Acts of Governmental Waste, Abuse, and Corruption*, 95th Congress, 2nd Session (February 1978).

U.S. Senate, Subcommittee on Health, of the Committee on Labor and Public Welfare, *Hearings on Human Experimentation*, 93rd Congress, 1st Session (February 21,22; February 23, March 6; March 7,8; April 30; June 28,29; July 10, 1973).

U.S. Senate, Subcommittee on Separation of Powers, of the Committee on the Judiciary, *Hearings on the Human Life Bill* [*S.158*], 97th Congress, 1st Session (April 23,24; May 20,21; June 1,10,12,18, 1981).

U.S. Surgeon General, *The Health Consequences of Smoking* (U.S. Government Printing Office, Washington, D.C., 1982).

Vandenbergh, John G., ed., *Pheromones and Reproduction in Mammals* (Academic Press, New York, 1983).

Van den Haag, Ernest, "Against natural rights," *Policy Review,* no. 23: 143–175 (1983).

Vaupel, James W., "The prospects for saving lives: A policy analysis," printed as 44–199 of U.S. House of Representatives, Subcommittee on Science, Research and Technology, of the Committee on Science and Technology, *Hearings on Comparative Risk Assessment*, 96th Congress, 2nd Session (May 14–15, 1980).

Veatch, Robert M., "Justice and valuing lives," 147–160 of Steven E. Rhoads, ed., *Valuing Life: Policy Dilemmas* (Westview Press, Westview, Colo., 1980).

Veatch, Robert M., *A Theory of Medical Ethics* (Basic Books, New York, 1981).

Veyne, Paul, *Writing History*, translated by Mina Moore-Rinvolucri (Wesleyan University Press, Middletown, Conn.,1984).

Waddington, C. H., *The Ethical Animal* (Atheneum, New York, 1961).

Wade, Nicholas, "Sociobiology: Troubled birth of new discipline," *Science* 191: 1151–1155 (1976); reprinted as 325–332 of Arthur L. Caplan, ed., *The Sociobiology Debate* (Harper & Row, New York, 1978).

Wade, Nicholas, *The Ultimate Experiment: Man-Made Evolution* (Walker, New York, 1979).

Wade, Nicholas, "Food board's fat report hits fire," *Science* 209: 248–250 (1980).

Wanzer, Sidney H., S. James Abelstein, Ronald E. Cranford, Daniel D. Federman, Edward D. Hook, Charles G. Moertel, Peter Safar, Alan Stone, Helen B. Taussig, and Jan van Eys, "The physician's responsibility toward hopelessly ill patients," *New England Journal of Medicine* 310: 955–959 (1984).

Watson, James D., and John Tooze, *The DNA Story: A Documentary History of Gene Cloning* (W. H. Freeman, San Francisco, Calif., 1981).

Wecter, Dixon, F. O. Matthiessen, Detlev W. Bronk, Brand Blanshard, and George F. Thomas, *Changing Patterns in American Civilization* (University of Pennsylvania Press, Philadelphia, 1949).

Weil, Vivian, ed., *Beyond Whistleblowing: Defining Engineers' Responsibilities* (Center for the Study of Ethics in the Professions, Illinois Institute of Technology, Chicago, Ill. 60614, 1983).

Weinberg, Alvin, "Social institutions and nuclear energy," *Science* 177: 27–34 (1972a).

Weinberg, Alvin, "Science and trans-science," *Minerva* 10: 209–222 (1972b).

Weinberg, Alvin M., "The limits of science and trans-science," *Interdisciplinary Science Reviews* 2: 337–342 (1977).

Weinberg, Steven, "Reflections of a working scientist," *Daedalus* 103, no. 3: 33–46 (1974).

Weinstein, Milton C., and Harvey V. Fineberg, *Clinical Decision Analysis* (W. B. Saunders, Philadelphia, 1980).

Weinstein, Milton C., "Cost-effective priorities for cancer prevention," *Science* 221: 17–23 (1983).

Weiss, Carol H., "Research for policy's sake: The enlightenment function of social research," *Policy Analysis* 3: 531–545 (1977).

Wenz, Peter S., "Ethics, energy policy, and future generations," *Environmental Ethics* 5: 195–209 (1983).

Westin, Alan F., *Whistle-blowing: Loyalty and Dissent in the Corporation* (McGraw-Hill, New York, 1981).

Whitehead, Alfred North, *Science and the Modern World* (Macmillan, New York, 1925; reprinted by The Free Press, New York 1967).

Whittemore, Alice S., "Facts and values in risk analysis for environmental toxicology," *Risk Analysis* 3, no. 1: 23–33 (1983).

Wiegele, Thomas C., *Biology and the Social Sciences: An Emerging Revolution* (Westview Press, Boulder, Colo., 1982).

Wiener, Norbert, "A scientist rebels," *Atlantic Monthly* 179, no. 1: 46 (1947); also printed in *Bulletin of the Atomic Scientists* 3, no. 1: 31 (1947).

Wiener, Norbert, "A rebellious scientist after two years," *Bulletin of the Atomic Scientists* 4: 338 (1948).

Wiener, Norbert, *I Am a Mathematician* (MIT Press, Cambridge, Mass., 1956).

Wiesner, Jerome B., *Where Science and Politics Meet* (McGraw-Hill, New York, 1961).

Wilder, Russell M., "A brief history of the enrichment of flour and bread," *Journal of the American Medical Association* 162: 1539–1541 (1956).

Wilford, John Noble, *The Mapmakers* (Alfred A. Knopf, New York, 1981).

Williams, Robin M., Jr., *American Society* (Alfred A. Knopf, New York, 1970).

Wilson, Carroll L., "Nuclear energy: What went wrong," *Bulletin of the Atomic Scientists* 35, no. 6: 13–17 (1979).

Wilson, Edward O., *Sociobiology: The New Synthesis* (Harvard University Press, Cambridge, Mass., 1975a).

Wilson, Edward O., "Human decency is animal," *New York Times Magazine*: 39ff. (October 12, 1975b).

Wilson, Edward O., "Academic vigilantism and the political significance of sociobiology," *BioScience* 26, no. 3: 183ff. (March 1976); reprinted as 291–303 of Arthur L. Caplan, ed., *The Sociobiology Debate: Readings on Ethical and Scientific Issues* (Harper & Row, New York, 1978).

Wilson, Edward O., *On Human Nature* (Harvard University Press, Cambridge, Mass., 1978).

Wilson, Robert R., "The conscience of a physicist," *Bulletin of the Atomic Scientists* 26, no. 6: 30–35 (1970).

Winner, Langdon, *Autonomous Technology: Technics-Out-of-Control as a Theme in Political Thought* (MIT Press, Cambridge, Mass., 1977).

Winner, Langdon, "Do artifacts have politics?" *Daedalus* 109, no. 1: 121–136 (1980).

Wolff, Mary S., "Occupationally derived chemicals in breast milk," *American Journal of Industrial Medicine* 4: 259–281 (1983).

Wright, Christopher, "Scientists and the establishment of science affairs," 257–302 of Robert Gilpin and Christopher Wright, eds., *Scientists and National Policy-making* (Columbia University Press, New York, 1964).

Wulff, Keith M., ed., *Regulation of Scientific Inquiry: Societal Concerns with Research* (Westview Press, Boulder, Colo., 1979).

Wynne, Brian, *Rationality and Ritual: The Windscale Inquiry and Nuclear Decisions in Britain* (The British Society for the History of Science, Halfpenny Furze, Mill Lane, Chalfont St. Giles, Bucks HP8 4NR, 1982).

Yalow, Rosalyn, "Fear of radiation," op-ed essay, *New York Times*: 22 (January 31, 1979).

York, Herbert, *The Advisors: Oppenheimer, Teller, and the Superbomb* (W. H. Freeman, San Francisco, 1976).

Ziman, John, *Public Knowledge: The Social Dimension of Science* (Cambridge University Press, Cambridge, 1968).

Zimbardo, Philip G., "On the ethics of intervention in human psychological research: with special reference to the Stanford prison study," *Cognition* 2: 243–256 (1973).

Zinder, Norton D., "The Berg letter: A statement of conscience, not of conviction," *Hastings Center Report* 10, no. 5: 14–15 (1980).

Zuckerman, Solly, *Scientists and War: The Impact of Science on Military and Civil Affairs* (Harper & Row, New York, 1967).

Index